Critical Muslim 31

Climate

Critical Muslim is published quarterly by C. Hurst & Co. (Publishers) Ltd. on behalf of and in conjunction with Critical Muslim Ltd. and the Muslim Institute, London.

All editorial correspondence to Muslim Institute, CAN Mezzanine, 49–51 East Road, London N1 6AH, United Kingdom.
E-mail: editorial@criticalmuslim.com

The editors do not necessarily agree with the opinions expressed by the contributors. We reserve the right to make such editorial changes as may be necessary to make submissions to *Critical Muslim* suitable for publication.

C. Hurst & Co (Publishers) Ltd., 41 Great Russell Street, London WC1B 3PL

ISBN: 978-1-78738-218-3 ISSN: 2048-8475

To subscribe or place an order by credit/debit card or cheque (pounds sterling only) please contact Kathleen May at the Hurst address above or e-mail kathleen@hurstpub.co.uk

Tel: 020 7255 2201

A one-year subscription, inclusive of postage (four issues), costs £50 (UK), £65 (Europe) and £75 (rest of the world), this includes full access to the *Critical Muslim* series and archive online. Digital only subscription is £3.30 per month.

A Cataloguing-in-Publication data record for this book is available from the British Library

Critical Muslim

Subscribe to Critical Muslim

Now in its eighth year in print, *Critical Muslim* is also available online. Users can access the site for just £3.30 per month – or for those with a print subscription it is included as part of the package. In return, you'll get access to everything in the series (including our entire archive), and a clean, accessible reading experience for desktop computers and handheld devices — entirely free of advertising.

Full subscription

The print edition of *Critical Muslim* is published quarterly in January, April, July and October. As a subscriber to the print edition, you'll receive new issues directly to your door, as well as full access to our digital archive.

United Kingdom £50/year

Europe £65/year

Rest of the World £75/year

Digital Only

Immediate online access to *Critical Muslim*

Browse the full *Critical Muslim* archive

Cancel any time

£3.30 per month

www.criticalmuslim.io

CM31
July–September 2019

CONTENTS

CLIMATE

CLIMATE

INTRODUCTION: ENDGAME

Ehsan Masood

There is a small box-shaped hedge at the front of the house where I live in the southeast of England. It's adjacent to the main front door and measures about a couple of cubic metres. It's the last thing I see when I leave home to go out to work, and the first thing I see when I return. The hedge has a reassuring quality about it: it's an immovable, dependable, ever-watchful biological gatekeeper.

Three days before finishing this essay introducing the climate issue of *Critical Muslim*, I glanced at the hedge as I usually do when leaving the house. In spite of intermittent rains, it seemed unusually off-colour. On closer inspection I realised that the leaves, which were formerly green, had now turned a greyish-brown, and were joined together in a haze of webbing. I peered more closely and could see that every twig was infested with a species of caterpillar. There were in fact hundreds of caterpillars, moving and rolling very, very slowly, as if at the end of a particularly heavy Iftar or Christmas dinner. This was the box tree caterpillar and, as I later discovered, it is on a mission to defoliate gardens up and down the British Isles. The box tree caterpillar is remarkably efficient: it does its job in forty-eight hours. That's forty-eight hours to destroy the leaves on a hedge that, I am reliably told by elderly neighbours, has stood in its place for at least fifty years.

The box tree caterpillar is thought to have journeyed to the UK from east Asia twelve years ago. Its mode of arrival might have been via the beak of a migratory bird, or, perhaps the checked baggage of an unwitting air traveller. So far, there's no targeted remedy. The august Royal Horticultural Society says the safest method of control is to remove the caterpillars by hand. But as this is not very practical, most gardeners I suspect are doing as I did and resorting to some variety of insecticide. But in taking this one, infinitesimally small step for mankind, I simultaneously took a giant leap towards the coming mass extinction.

Yes, a *mass extinction*, as in *extinction* of human, animal and plant life on a *mass* scale is coming. That's the bad news. The better news is that it won't happen in our own lifetimes; nor in our children's, nor in their children's lifetimes; and nor indeed for several more generations to come. A mass extinction is, thankfully, a long way off, except that the signs are unmistakable and the majority of humanity is singularly unprepared for it.

We know there's a mass extinction coming because conservation biologists have been repeating for at least the past two decades that the rate at which species are becoming extinct is higher now than at any time since the last (fifth) mass extinction. This is when plant and animal life was practically wiped out around 65 million (ish) years ago. Whereas the previous mass extinction was out of our hands – the leading cause is understood to have been a meteorite hitting Earth – the next one will be entirely of our own doing.

Extinction is natural – not all species can survive or move to new environments if their survival is at stake. At the same time, not all extinction is bad, and some extinction – like, I would argue, the box tree caterpillar for example – is desirable. More seriously, no one wants polio or smallpox hanging around. We all want to see eradication, or at least better control, of the mosquito that causes malaria. But our current extinction rates are way more than what we might expect them to be – 100 times the background – maybe more. And they are showing no sign of going backwards. So we may be the last human species – at least for a while.

But let's suppose we survive the coming mass extinction. Or imagine for a moment that all those biologists will need to tweak their extinction projections, most likely from errors in their estimates of time scales. It could be that humanity will still be around for another million years or so. Even so, there's another existential threat coming down the slipway, and that is climate change. Thanks to our addiction to carbon, to coal and to natural gas, global average temperatures are climbing, and the effects are already being felt in the form of more weather extremes. Dry regions are getting drier; wet regions are wetter. Melting ice at the poles is adding to sea level rise, which in turn is threatening the survival of small islands, especially those in the southern hemisphere nearer the Antarctic. Farmers are slowly losing land that they once used for crops and for pasture. The consensus of those researchers who belong to the Intergovernmental Panel on Climate Change (IPCC), the body that advises the United Nations on climate science, is that we could have less than fifteen years to turn things around.

The likelihood of such a turnaround is an open question. We're talking about fifteen years for electric vehicles to replace every single petrol and diesel scooter, car and truck; fifteen years for solar, wind and wave energy to replace coal-fired power plants. Could it happen? The optimist in me says, yes it's possible, but the realist says it's a very, very big ask – and perhaps too big.

There are at least FIVE big problems to making an appreciable dent on both species loss and on climate change. Let's deal with each in turn.

Problem ONE: climate policies need equity

The challenge of dealing with climate change is not just about combating the scale and the complexity of a question such as how does one get every household on the planet to stop using gas to cook, and to heat their homes. This particular question assumes that every household already has piped natural gas. But many don't and nor do they have reliable electricity, nor running water. Often, such households live in unstable and insecure accommodation. And when it comes to climate change policy-making, their voice is still largely missing. In the rush to declare climate emergencies all over the world, there are tens of millions of people who have waited for the fruits of economic growth to reach them, and they are at genuine risk of having what is owed to them snatched a the last minute.

In Muhammad Akbar Notezai's article, 'Little London', we meet Abdur Rehman, a pastoralist in his seventies from Pakistan's western province of Balochistan. Balochistan is in parts green and in parts arid. Over the years, the green parts have been shrinking, as the arid zones expand and there's a strong likelihood that this is happening as the effects of climate change start to bite. Average temperatures have risen four degrees in a decade and in that time, probably before, Abdur Rehman has lost his land to drought and watched his goats die, one-by-one because of a shortage of water.

The paradox is that the situation he finds himself in would have been the same as that of a pastoralist in Europe a century ago. Abdur Rehman's Victorian forebears also lived off the land; they lacked security in housing and in employment; and they had little if any welfare, nor pensions protection. Industrialisation didn't only make them better off, but also better prepared them for the effects of climate change. But Instead of preparing for retirement, as his counterparts in Europe are now doing, Abdur Rehman tells our reporter Muhammad Akbar Notezai that he is preparing to die.

While there are still swathes of humanity in the developing countries who have yet to see electricity, or who are waiting patiently to own or travel in motorised transport, they – and their governments – will have little tolerance for any argument that says they must continue to live like it's still 1719.

The richer countries certainly seem that they are on an unstoppable path to decarbonising, and this is thanks in no small measure to the efforts of schoolchildren led by Swedish teen Greta Thunberg and because of the passionate and honest commitment of Extinction Rebellion members such as James Brooks, who describes his experiences with the organisation in his essay 'In Love and Rage'. But there is as yet little prospect – and certainly no equity – in expecting the people of poorer countries to follow suit if it means having to forego their right to a path out of poverty. That question of equity is a deep dilemma confronting those advocating tackling climate change, and it will need every milligram of effort, initiative, ideas, inspiration, non-violent direct action and, frankly, a few miracles, if it is to be solved.

Problem TWO: We need to find a different way to measure growth

In addition to our addiction to fossil energy, humanity is guilty of another addiction too, which – at least in the richer countries – is getting in the way of taking meaningful action on climate change and species loss: this is our addiction to how we measure economic growth.

Governments and businesses have known since at least the late 1960s that there are damaging consequences to injecting soils with synthetic chemicals; and that there are damaging consequences to flooding the atmosphere with greenhouse gases such as carbon dioxide and methane. By the early 1970s, there was enough of an international scientific consensus that we have a problem that needs looking at. In response, a number of world leaders – including India's then prime minister, Indira Gandhi – gathered for the first UN Conference on Environment and Development in Stockholm. At the end of a week's tough negotiations the assembled leaders promised to take steps to ensure that their future industrial policies would be more environmentally responsible. This was for its time a superlative achievement as the majority of countries resisted the idea that human activities could be causing irreversible climate change.

In the ensuing five decades, there have been thousands of new laws, rules and regulations that protect human and planetary health. They include myriad measures from regulating pesticide residues in fruit and vegetables, to legislating the size and volume of pollutants spewing forth from diesel engines – not to mention the 2015 Paris agreement on climate change. But as we know all too well, a majority of all the laws and the rules have failed in two respects: they have failed to limit climate change; and they have failed to slow down, never mind halt, species extinction. Indeed, I would argue that this is a failure partly by design.

How so? Well, as I've argued elsewhere in my book *The Great Invention*, in order to genuinely slow down the rate of climate change, or to slow down species extinction, we have to control the conversion of forest land for agriculture, homes or industry. These are not only difficult goals for countries which still have large populations in poverty, but they are also hard for industrialised economies too.

The problem is in how we measure and how we celebrate economic growth. Growth is measured using the measurement known as Gross Domestic Product (or GDP). Every quarter, when a country's finance ministry announces economic growth rates of 2 or 4 or 6 per cent, it's like a celebration: a celebration of more government spending and a sign that more of us are spending more time buying things in the shops. But in a world where energy is powered by carbon, higher economic growth also means more carbon dioxide in the atmosphere and continued loss of species on the ground. And that is why GDP is so problematic for efforts to tackle climate change and species loss. It's kind of chicken-and-egg: our governments are locked into measuring the worth of their economies through GDP – and that means GDP always needs to rise. But GDP's continued rise relies on carbon-based development. So if we cannot quickly change the carbon-based trajectory of our economies, we can at least find a more inclusive measure.

Problem THREE: How to stop Big Oil's addiction to money

I used to think there was some great mysterious art to the kinds of eye-watering profits that large corporations are able to make, but that was before I started looking more closely at the extractive industries. One of the many reasons why the great powers were in an unseemly haste to profit from the discovery of some of the earliest oilfields at Dammam in Saudi Arabia at the start of the twentieth century, is because oil is an astonishingly profitable investment. The biggest requirement is capital, but once that's secured, the path to riches is pretty much guaranteed. And if you happen to be a member of one of the ruling families of the Middle Eastern economies, it means you need to do little more than sit back and watch the creation of dollars out of thin air – what the economist Jeffrey Sachs of Columbia University in New York famously coined the 'Resource Curse'.

When countries and companies find a surefire way to profits, what tends to happen is that they don't like to give it up easily. Not only that, but there's lots of evidence that they will play dirty tricks against anyone – not only competitors – who might want to interfere with the tap from which flows the liquid gold. That fundamentally lies at the heart of the approach taken by members of the Organisation of Petroleum Exporting Countries (OPEC) and their partners in the oil and gas industries, against those arguing that extracting carbon from the soils and then pumping it into the atmosphere isn't a good thing for the climate.

For a few years in the mid-to-late 1990s, it was my job to report from international climate change conferences. This was the period running up to the 1997 Kyoto Climate Protocol, the first legally-binding agreement in which the richest countries agreed to make modest cuts to their greenhouse gas emissions. The protocol was agreed two years after an international team of scientists working for the IPCC had established beyond doubt that human activities such as burning fossil fuels, are warming the planet.

Even today, international climate change meetings take place in cavernous conference centres, or in very large hotels because of the need to accommodate sometimes thousands of delegates. They include the representatives from individual countries, but their numbers are often far exceeded by special interest groups, such as environmental organisations like Greenpeace, by scientists, and also representatives from industry. By industry, I mostly mean the oil and gas giants including Shell and Exxon Mobil, which for many years were members of a lobby group that went by the unlikely name of the Global Climate Coalition (GCC).

The actual climate negotiations, the talks themselves, happen at the level of governments, in that government representatives meet with each other in a parliament or UN-style seating arrangement. Ideas are proposed, debated, rejected or agreed upon by a consensus of those present. Lobby groups can in some instances observe what is happening, and can of course meet government delegates outside of the main meetings, but they are not supposed to interfere with proceedings. This, however, is not how the GCC chose to operate. Its representatives were openly advising OPEC countries on the wording of proposed climate agreements, and, time after time, GCC staff made no secret of their attempts to water down language that attributed climate change to human activities.

Paradoxically, this was one of those rare occasions where the delegates from Saudi Arabia were pally with their colleagues from Iran; and the delegates from Iran could be seen sharing jokes with delegates from the US as well as with oil company representatives. It's not difficult to see why this was happening: each had a shared objective, and that was to obstruct at all costs any effort or activity that would interfere with oil and gas production.

Problem FOUR: The uses and abuses of religion

Once, during one such international climate change conference I had the pleasure of interviewing Saudi Arabia's chief climate negotiator and I remember asking him something to the effect that, 'if you come from an Islamic state, wouldn't you agree that God wants us all to protect nature?' He paused before replying: 'That is religion. This is politics'.

Now, nearly twenty-five years after that encounter, and with the Global Climate Coalition consigned to history, the OPEC countries are falling over themselves to show that they, too, are looking to a post-carbon future – and some are displaying their new-found environmentalism with an explicitly Islamic badge. A decade ago, the rulers of Abu Dhabi planned to create an entirely zero-carbon city, called Masdar. Qatar, in contrast, created a more modest 'Qur'anic Botanic Garden'. This combines architectural elements of Islamic-era gardens such as fountains and walkways, with plants that are mentioned in the Qur'an and hadith. Not to be outdone, Dubai has its own version, and bigger, as you would expect, which opened in March 2019.

At the global level, there are initiatives such as the Islamic Declaration on Climate Change, which call on those in power to 'tackle habits, mindsets, and the root causes of climate change, environmental degradation, and the loss of biodiversity in their particular spheres of influence, following the example of the Prophet Muhammad (peace and blessings be upon him), and bring about a resolution to the challenges that now face us.' And nationally, there's a growing movement of religions in the service of combating climate, such as Faith for the Climate Network.

In theory, it ought not be difficult to be an Islamic environmentalist, because Islam has an established stewardship ethic. In common with many, I, too, was taught from an early age to regard my presence on Earth as a temporary trusteeship, to be handed on to the next generation. This I understood to mean that humans need to tread lightly, to use resources carefully and to consider public interest, equity and the needs of future generations when allocating resources.

Perhaps another reason why Islamic environmentalism has theoretically less road to travel is the absence of a strong tradition of anthropocentrism – although Islam is not entirely free from this phenomenon.

Classical Islamic-era astronomers, for example, had no problem in accepting that there could be a Sun-centred Universe. The eleventh-century Spanish Sufi mystic Ibn Arabi developed an idea that he called the Unity of Existence or Unity of Being in which he suggested that God and creation share many of the same characteristics. The historian of Sufism Mark Sedgwick has argued that Sufism could have been Islam's response to the profligacy of the second and subsequent generations of Caliphs, that it was a reaction to the ostentatious lifestyles of the rich, a reaction to despotism, as well as waste.

In contrast, as Giles Goddard notes in 'Too Close to the Sun', five decades ago, Lynn White Jr., an American historian of the Middle Ages, attempted an explanation of the relationship between Christianity and the environment in an essay in the journal *Science*. The essay was called: 'The historical roots of our ecological crisis' and in it, White Jr. explained how a strand within Christianity believes that humans are permitted to benefit from the natural world because they occupy creation's number-one spot. Evidence for what has come to be known as the Dominion hypothesis, including this from Genesis 1: 28:

> Then God said, 'Let us make humankind, in our image, according to our likeness; and let them have dominion over the fish of the sea, and over the birds of the air, and over the cattle, and over all the wild animals of the earth, and over every creeping thing that creeps upon the earth.

White's paper caused something of a firestorm at the time, but much has changed since its publication in 1967. There has been an emergence – a re-emergence – of a strong tradition of Christian environmentalism, which also draws on scripture, but to argue the opposite case: that humans are God's stewards – that we do not own nature, but are guardians, trustees, even tenants.

One of the first among all of the faith groups to support the scientific consensus on global warming was the World Council of Churches, and they have had an official presence at UN climate talks from the 1980s. Moreover, prominent scientists such as John Houghton, a former head of the UK Met Office and a leading member of the Intergovernmental Panel on Climate Change, have also played their part in organising conferences on both sides of the Atlantic to persuade more sceptical US evangelical communities to take climate change more seriously.

In contrast, and rhetoric aside, Islam's environmentalists have further to travel. The stewardship ethic exemplified in the life of the Prophet Muhammad, and the Sufi ideal of self-denial and asceticism is light years from the environmental record of today's Muslim world. Yale University in the US produces a list of every country, ranked according to what it calls an Environmental Performance Index. Countries are scored by a number of indicators such as air quality, water resources, biodiversity and habitat protection and the development of clean and sustainable energy sources.

Each of these scores is then compressed into a single number–an index. The top score being 100 and the lowest being zero.

In the tables for 2018, the best performing Islamic country is Qatar which is thirty-second overall in the world. There are no Muslim-majority countries in the global top ten. Or the top twenty. Only Qatar is in the top fifty. Around half of all countries in the Organisation of Islamic Countries network are ranked 100 or below. Iran is eightieth; Saudi Arabia is eighty-sixth; Indonesia is ranked 133rd. Many of these countries are below where they were on the index a decade ago.

Problem FIVE: Big powers like technology

Industrialisation began centuries before the industrial revolution: it began when the first farmers discovered they could grow crops by planting seeds. This was as big for its time as the invention much later of electricity, wireless telecoms, and then the Internet. Instead of roaming the landscape in search of food, the discovery of agriculture meant that human populations could settle in one place and start to build communities. They could grow their own food, but also sell any surplus. Farming was in that sense a step up, a means to a higher standard of living, a path to the aspirational middle class, and agriculture soon spread throughout the world.

For most of those 12,000 years, crops were grown using largely natural methods, and on a relatively low scale in that the majority of farmers would grow food to feed themselves and their families, and then sell any surplus at the local market. What changed farming irrevocably is the introduction of industrial-scale technologies, which, in turn, seeded the birth of agri-businesses. The combine harvester did away with employing slow and expensive humans at harvest time. Modern chemistry leant a helping hand as agrochemicals killed pests and weeds, and were used as fertilisers and preservatives. All of this allowed crops to be grown at scale and shipped to markets thousands of miles from where they were grown.

There is of course much more to industrialisation than farming. Two of the greatest discoveries that have brought prosperity to billions are electricity and the internal combustion engine and centuries after their discovery, they remain a valuable and sought-after commodity. Every nation – and perhaps every citizen in every nation – wants and needs power and motorised transport access to these and other technologies. Electricity extends daylight hours well beyond sunset. It keeps us warm in the winter and cool in summer. Electricity powered to engines allows us to travel greater distances for work and increasingly for leisure too. And now electricity allows us to communicate in previously unimaginable ways.

Much of the science and technology that we take for granted today emerged during the age of Europe's great empires: and that includes electricity, navigation, the steam engine, telecommunications. The second voyage of HMS Beagle, which had on board one Charles Darwin, would not have happened without the financial clout and military protection from a serious naval power as Britain was at the time. James Cook, the naval captain who claimed the 'discovery'" of Australia two centuries ago was charting a more explicitly scientific voyage on behalf of Britain's elite science academy, the Royal Society.

One of my favourite non-fiction reads is sociologist Zaheer Baber's *Science of Empire*. Baber tells the story of how Britain brought modern science to India. Nearly a quarter century after it was first published, I still find myself dipping into tales of how Britain's scientific elite sought to replicate the knowledge institutions of Britain – some of which were themselves quite new – in India. Peppered through his accounts of the creation of botanical gardens, professional scientific societies and official scientific advisory boards are some gems of the English language, which provides flashes of insight into the motivations of those involved in the colonial project, such as this from Sir Charles Trevelyan:

> The peculiar wonder of the Hindu system is not that it contains so much or so little of true knowledge, but that it has been so skilfully contrived to arrest the progress of the human mind. Our duty is not to teach, but to unteach.

Islam's empires similarly gave rise to advanced science and technologies including irrigation, mapping, surgery and medical instrumentation, along with basic science, philosophy and the mathematics that underpins so much technology – notably algebra. Some rulers, notably the Abbasid Caliph Al-Mamun (r. 813–833), approached belief in a similar way to how they approached science – through argument, evidence, experiment, and, yes, through cruelty and despotism, just as Europe's colonists would a century later. Mamun was famously an arch-rationalist and wanted desperately for all in his kingdom to be rationalist too. Fortunately for Mamun he had the power to implement his will. Less fortunate were those who had no wish to bend to Mamun's world-view, and who suffered greatly as a result.

Technology allied to a (somewhat smaller degree) of despotism remains a characteristic of today's big powers too, and for all the noise about going carbon neutral, there is little prospect of the three big powers – the United States, China and the European Union – deliberately cutting back on industrial technology, if it means a diminution of their global status as superpowers.

The development and subsequent desire for modern technologies and our yearning for modern ways of living are of course not the only causes of the present crisis in global warming and species loss, but they are a very significant contributory factor. It is undeniable that our current malaise can be attributed in parts to unchecked industrial growth, to science and technology in the service of empires, and to leaders with a highly instrumentalist approach to governance and to belief.

All things considered, this is ultimately what makes the task of pulling humanity from the brink of the abyss so very, very difficult. That said, we have to live in hope, but it is not much more than a faint flicker of a shard of hope, that humanity can pull itself out of the deep hole we have dug for ourselves.

THE *BARAKAH* OF WATER

Medina Tenour Whiteman

'We created every living thing from water'.
Qur'an 24:45

Imagine a valley where orange trees nestle among gnarled, millennia-old olives; where figs, mulberries and pomegranates hang heavy with juice; where almonds are studded with nutty goodness in summer and burst with delicate pink-white flowers in early spring; and where channels of snowmelt ribbon their way from majestic peaks around mountain flanks, rushing into smallholdings and keeping the land alive.

One sunny autumnal afternoon, wearing a *kufi* hat and a long grey beard, Abu Bakr hoes a plot of land, turning over clods bright with yellow wood sorrel flowers to sow broad beans, cabbages, cauliflowers and pumpkins. Chickens strut about officiously. Bees buzz. A few sheep baa behind a fence. From the mosque on the land below, the *adhan* for *'asr* can be heard.

You'd be forgiven for thinking this was a page from a history book from some faraway Middle Eastern land, but this is Spain, in the twenty-first century. Welcome to the Alpujarra mountains, halfway between Granada and the Mediterranean Sea.

This area was practically untouched when, about 1,000 years ago, Andalusi Muslims began cultivating the land in ways that would still define traditional agriculture today. They hand dug terraces into steep slopes so that the rain wouldn't wash the earth away into the river below. Then they planted trees on these terraces and watered them with a sophisticated network of irrigation channels: *as-saqiyyat,* Hispanicised as *acequias*. It's one of thousands of Arabic words that have remained in the Spanish language, among them the Qur'anic word for a seed, *habba* – related to *hubb* (love) – a word that has survived as *haba*: the humble broad bean.

Órgiva, the hub of the Alpujarras, was known to the Greeks and described by the Arab chronicler Al-Udri in the eleventh century. In the last few decades, it has become a haven for alternative communities fleeing rat races all over the world, seeking health in body and mind and the chance to raise their families in harmony with nature. Among these newcomers are a diverse bunch of Muslims, mainly with Sufi leanings, who share the concerns of their multinational neighbours.

Abu Bakr is a Spanish convert who has been farming near Orgiva for twenty-five years. 'When I first came here, the climate was very different,' he says. 'The winters weren't so stormy, or the summers so hot.' In spite of the dam that was built in 2004, to which locals attribute the muggy summers and mosquitoes, the land is becoming gradually drier. The peaks of the Sierra Nevada used to be covered in snow all year round, but now it's dry for months during the summer. It might seem strange to think of mountains like these as desert, with their wild rosemary, lavender, and yellow broom on slopes too steep or dry to cultivate, but land doesn't need to look like the Sahara to be desertified.

Sustainable farming is a complex balance between many factors: respecting the soil, allowing it to rest at times and replenishing nutrients with manure and green fertilisers; sowing a range of plants that support one another; and careful animal husbandry, to name but a few. Andalusis were master agriculturalists: between the tenth and fourteenth centuries they wrote a huge number of treatises, which included Khayr al-Din ibn Ilyas's *Kitab al-fallah* (The Successful Farmer) and the *Kutub al-Filaha* (Books on Husbandry) – canonic texts for an 'Islamic Green Revolution'.

In a place where rains can be absent for months and then come in torrents that cause landslides, the key to this delicate balance is water management. This sophisticated skill, borne of the dry land farming of Arab lands, created a veritable Eden in the Emirate of Granada – home to about a million people at its zenith – and has been passed down over generations to present times. The villages of the Alpujarras were all built around springs, for instance the spa town Lanjarón, whose name is probably derived from *'Aynu Harun*: Haroun's Spring. The same story is found all over Andalusia: irrigation disputes are still adjudicated on the steps of the main church in Valencia, as they were in the Islamic period, when it was a mosque.

From 1567 to 1616, a group of Moriscos rebelled against the new political order. These Spanish Muslims, whose forced conversions to Catholicism were always suspect, were rounded up and exiled from the Granada Emirate, dispersed all over Spain – if they survived the weeks of travel on foot. As many as a million were eventually deported, mainly to North Africa. But a few families remained: one to tend the mulberry groves, the source of Granada's lucrative silk trade (silkworms can only feed on white mulberry leaves), a few to preserve trades such as shepherding and blacksmithing, and one to maintain the *acequias*.

This was essential, as the 'Old Christians' who were brought from the north of Spain to repopulate this once-thriving area didn't have a clue how to cultivate food in this often harsh mountain environment (geologically, the Sierra Nevada is related to the Rif mountains of Morocco). So the Muslims' knowledge of land stewardship didn't disappear, even though all memory of the Muslim period was suppressed. Only a few years ago, an elderly local advised his neighbour to plant her potatoes 'facing Mecca'.

Narrow strips of terraced land aren't only a clever way of resisting erosion, they also lend themselves to varied smallholding agriculture rather than the economically-driven monoculture of the coast, which is blighted by a sea of greenhouses. These are an ecological nightmare in so many ways: the plastic is usually dumped when it wears out; plants are forced to grow out of season, boosted with artificial fertilisers (think of that next time you buy fresh tomatoes in January); and the staff are, more often than not, immigrants from North and West Africa who have no papers, and are therefore easily exploited – not to mention exposed to the noxious products being used. Yet when these chemicals are taken out of the equation, the produce can still be labelled 'Organic', even though the other systemic problems such as soil degradation and air freighting remain.

It's no wonder, then, that many people are turning to sustainable traditional agriculture, or the modernised version presented by Permaculture. As local teacher Ras John Cresswell of Supernatural Permaculture explained to me, it is founded on a combination of Earth Care and People Care, enhancing soil fertility for future generations rather than exploiting the earth for short-term gains. Permaculture seeks to update ancient, holistic farming techniques with technological and scientific approaches, 'regreening' degraded terrains – with some phenomenal success stories.

My everyday Permaculture gurus are usually my neighbours, retired Spaniards who have turned their hands to small-scale horticulture and who are so generous in sharing their gluts of fruits and vegetables, seeds, cuttings, fruit tree grafts, and – especially – information on when to sow (always related to saint's days, the moon phases, or the seasons). 'When there's enough, there's enough for everyone', as my neighbour Paco says. Even people living in town like to keep boxes of onions, lettuces and broad beans on their rooftops; self-sufficiency is a habit too deeply-ingrained to kick.

The *acequias* are the veins of these mountains. Granada's *acequias* run for twelve miles, carrying water to the city as well as the Alhambra and the farmlands surrounding the old Zirid city walls. These channels ran down the middle of the Albaicín's cobbled paths, leading to cisterns or *aljibes* – vaulted underground chambers accessed via a hole at street level – that supplied the city's many public baths. No prizes for guessing that the word *aljibe* also comes from Arabic – *al-jubb* – or that they have long been important features of Middle Eastern water management strategies.

Acequias are a serious business. A 'community of irrigators' pays a minimal yearly fee for the use and upkeep of their *acequia*, and each one has a turn with a set number of hours depending on how many square metres of land they have. Woe betide you if you water your land outside of your time slot! It's caused more fisticuffs between farmers than you can shake an olive branch at. Standing ankle-deep in icy cold floodwater in the sweltering heat of summer, digging the earth with a hand hoe to let it reach every tree, moonlight glinting off the temporary ponds between the olives, is one of the most magical experiences I've had in the Alpujarras.

Flood irrigation has had such success in keeping this landscape abundant because it soaks the clayey, stony soil found here very thoroughly and deeply; even at the height of summer, when temperatures can soar well over 40C for weeks at a time, the soil stays damp for five days or longer, keeping thirsty fruit trees alive and plumping up green gems on the drought-resilient olives. Tony Milroy, an expert in arid farming and UK's advisor to Yemen's Ministry of Agriculture, points out that flood irrigation also benefits the microflora of the soil, as it brings with it seeds and micronutrients from elsewhere. Of course, the downside of this is if your neighbours upstream are using chemical fertilisers and weedkillers, as inevitably some of it will run off into the *acequia* – which can be like a miniature waterfall at full power.

To collect the many gallons of allotted *acequia* water when it arrives at silly o'clock in the morning, farmers often build an *alberca* (from the Arabic *al-birkah*, or pond, related to *barakah,* or blessing). While tube irrigation sounds like the most sensible way to use meagre water resources, installing a timer and laying black plastic tubes is beyond many people's means; besides, timers can break down, while the plastic of the tubes leaches hormone-like chemicals into the soil. It also encourages the use of weedkillers, as it's too awkward to take them up before ploughing the land – a favourite technique for removing weeds and making mechanical harvesting easier. Sprinkler systems add the issue of superficial water on leaves turning to magnifying lenses for the sun to literally burn them alive.

Today the *acequias* are still maintained by the same families that have cared for them for centuries, removing obstructions and rebuilding them when they're destroyed by heavy rains. Some have been cemented, which unfortunately deprives the land along the *acequia* from the water that seeps into it along the way; otherwise *acequias* are visible from a distance as a strip of greenery along the mountain slope.

But there are ways of preserving moisture that don't depend on *acequias*. A neat trick for a cottage garden is to sink an unglazed terracotta pot into the earth between plants, fill it with water and cover it, allowing the water to seep out slowly. Old boys in Lanjarón chop up prickly pear cactus leaves and dig them into the soil for time-released moisture. These cacti are also handy as firebreaks. Whenever I stumble across the veins of desiccated leaves I can't help but marvel at the exquisite beauty of their filigree mesh, like dry-land corals that once held the water alive in their leaves.

No matter how you manage water, the best way to keep it from evaporating is to make like a tree and (grow) leaves. Think of a forest, the thick layer of humus underfoot, the rich smell of decomposing leaves, and the dappled light filtering through the canopy. Forest farming aims to replicate these conditions, albeit maximising the edible output. I know what you're thinking: surely leaves need the sun for photosynthesis. True, but UV light destroys beneficial microorganisms in the soil that help plants extract nutrients from the soil; plus it can bake the earth to a hard, dusty crust that not even the most enthusiastic root system can break through.

Canopied tree farming designs have been used in oases in the Middle East for thousands of years. Simply put, tall, hardy trees like palms and olives offer shade to more vulnerable trees like oranges, which in turn protect grapevines, bushes, ground crops, and finally root crops. Of course, the Alpujarras aren't only exposed to the sun in summer, but below-zero temperatures in winter; this tiered system also helps protect plants from frost and wind.

At La Loma Viva, a nearby Permaculture project, they're pioneering a system called Syntropic farming, adapting it from its native Brazil to a semi-arid Mediterranean climate. In this method, young trees are planted surprisingly close together to mimic the natural environment of a forest, in rows to facilitate harvesting. Beneath the surface, the trees' roots share water, nutrients, even information about droughts and other phenomena to help one another to grow. Together with regular, heavy pruning, this leads to phenomenal growth. Their trees were planted as cuttings – figs, almonds, poplars, and the leguminous hardwood black locust – together with ground cover crops like chard, lettuces and artichokes, barely eighteen months ago. When I visited, the trees were taller than me – and they'd recently been pruned.

Yet Syntropic farming has bigger ambitions than merely producing edible abundance: it aims to reduce the need for irrigation to zero. What's more, it has been shown to create humid microclimates that bring dried rivers and springs back to life. The ground is heavily mulched with bark chips from their own chipper to prevent transpiration, keeping the soil damp. At the base of each tree a prickly pear cactus has been planted, which gathers dew and mist on its spines and lets the water trickle down to the earth.

Unusually, La Loma Viva doesn't have *acequias* but *qanat*s, tunnels leading into the mountain to reach hidden water sources – the mainstay of water management in arid environments like that of the Arabian Peninsula. In Yemen, it is the *qanat*s that ensure a water supply through the long rainless months, broken by a single monsoon season, during which the *wadis* suddenly burst into life. These downpours fill underground cavities in which the water is protected from evaporation, to be accessed via *qanats* during the rest of the year.

After a lifetime of traditional farming, Abu Bakr is retiring. A few younger Spanish Muslims are taking up the mantle, including Ishaq, who has a certified Organic farm near Lanjarón. But he's dubious about Permaculture: 'It requires a big investment and it can take years to see a return. What about the little guys in the meantime? A traditional family would manage with a hectare of olive and fruit trees, some goats, a *huerto* [vegetable patch], and they'd share their excess produce, seeds, cuttings and so on with their neighbours. We need the human element too. That's essential.'

Returning to subsistence farming isn't an attractive choice for many farmers today. It might feed a family, if you're lucky, but it isn't usually that lucrative – and without responsible management it can even lead to desertification. From Mali to Jordan to China, overgrazing has seen entire panoramas turn to dustbowls; livestock such as goats or sheep nibble the young saplings of pioneer trees that would have shaded other plants, thus creating an ecosystem.

The global agricultural tendency towards cash crops means that farmers have started giving preference to monoculture, in which one crop is grown over large plots of land. In the neighbouring province of Jaén, you can drive for hours and see nothing but rolling hills of young olive trees in immaculate rows; about half of all the world's olive oil is produced in Spain.

There are many reasons to dislike monoculture, among them the precariousness of relying on a single crop; if it fails, the farmer goes out of business – and, if it's an inedible crop like cotton, may even starve. The heavy machinery involved in ploughing, sowing and harvesting compacts the soil, reducing fertility. Besides, plants don't naturally grow in huge swathes of single species; many beneficial plant associations have been observed, in which one species protects another from a certain parasite or mould. For instance, Tagetes calendulas are thought to help protect Solanaceas from pests and nemotodes (threadworms).

The key problem with monoculture, however, is that it's motivated by financial gain rather than care for the ecosystem of which we are an integral part. Is there any better mirror for the state of our souls than the slash-and-burn devastation of old-growth forest to make a quick buck, only to find that the land turns to dust within a generation? As Karen De Vries from La Loma Viva says, 'The climate thing is just a symptom. What we need is a change in the climate of the mind.' And that begins with seeking deeper nourishment than merely food.

For Muslims it's not just an ecological no-brainer: Islam gives us the extra prerogative of humans being the *khulafa'* (viceregents, deputies) of Allah on the earth. The 2017 and 2018 editions of the Zawiyah Retreat in Alquería de Los Rosales, an Islamic conference centre based in the Segura mountains north of Granada, were called 'Tending the Earth: the Art of Living with God's Creation'. Many Permaculture practitioners and teachers are Muslims; there is even a Permaculture Research Institute in Hadramawt, Yemen, on land donated by the Ba 'Alawi Shaykh Habib 'Umar, who is reported to have said 'Permaculture is *wajib* [a duty]'.

There are so many parallels between the outer work of cultivating the land and the inner work of cultivating the soul. On a recent trip to Órgiva, the Qadiri Shaykh Muhammad Hydara al-Jilani from Gambia described this with marvellous simplicity: 'Farming is a path to God. Nature is a path to God. All paths lead to Allah.' Tilling the earth, sowing, watering, and harvesting all bear fruits in patience, effort, gratitude, and sharing.

Water is the key to this vitality. *Alberca* is related to *barakah* because the Arabic verb *b-r-k* means to kneel, the way a camel kneels to drink at a watering hole, or a person kneels to invoke a blessing in prayer. We need to be that close to the earth, so close we can see our own faces reflected in these blessed pools.

TOO CLOSE TO THE SUN?

Giles Goddard

It starts, perhaps, with Icarus. The boy and his father, Daedalus, are trying to escape from the island of Crete. Daedalus, the inventor, builds golden wings out of feathers and wax and straps them to himself and to his son. Before they take flight, the father issues a warning: my son, fly neither too low or the sea will dampen the feathers, nor too high or the sun will melt the wax.

Icarus takes flight, and is caught up in wonder at the world he sees, and at his own ability to fly. With a rush of delight, he rises, higher and higher, closer to the sun – and by the time he realises that the wax has indeed melted and every flap of his wings releases a cloud of feathers into the air around him, it is too late. His father sees him fall, and hears the splash. No more Icarus.

Or perhaps it starts with Prometheus, the trickster and challenger of the gods. In the myth as told by the poet Hesiod, he tricked Zeus into eating dry bones. In anger, Zeus hid fire from humanity: but Prometheus stole the fire in a fennel stalk, and brought it to earth. Zeus, in his wrath, condemned Prometheus to be tied to a rock, where every day an eagle came and plucked out his liver, and every night the liver grew back to be plucked out again. But the damage was done: humanity had fire, and could compete with the gods – and Prometheus was, in the end, rescued by Hercules.

Humanity faces an environmental crisis. The prospect of ecological catastrophe is real and imminent. Not only in relation to climate change; the Anthropocene age is producing an unprecedented level of species extinction, top soil is losing its fertility at an extreme rate, and the sea, as we know all too well, is becoming clogged by plastic. All the direct consequence of human activity.

How did it go so wrong? Is the human story that of Prometheus, stealing fire, transforming the world, and ultimately living to tell the tale, or that of Icarus, crashing flightless to a watery grave?

It is possible to trace a journey, step by step, which has led us to this point, a journey which has by the law of unintended consequences placed humanity in an unprecedented position where we have the potential to bring about our own extinction. My question is, whether we are seeing the outworking of the human capacity for pride, whether we have built ourselves a castle of hubris, whether we have overreached ourselves – or whether we can, like Prometheus, free ourselves from the punishment of the gods and survive to tell again a tale of hope and human flourishing.

The extraordinary development of knowledge and philosophy in the Muslim world between 800 and 1300 CE is well documented, as is the interaction between Christian and Muslim civilisations: the court of King Roger II of Sicily (1095–1154) featured large in my A level study of mediaeval history, his court a place of discovery and encounter, of tolerance for a range of religions, of scientific experiment and philosophical development. The first university in the West in a recognisable modern form was founded in Bologna in 1088 CE, evolving out of the schools for clergy which already existed and dedicated to the pursuit of scholarship and understanding.

The human drive towards enquiry and understanding – the Promethean theft of fire – had its outworking in courts and lecture halls around the Mediterranean – and at the pinnacle of the quest for knowledge was theology, the queen of the sciences. The greatest scholars of the period – Ibn Rushd, Maimonides, Roger Grosseteste – worked across the fields of philosophy, physics, medicine – indeed, the disciplines were at that stage barely separated. Grosseteste (1175–1253), who ended his life as Bishop of Lincoln, was well known and admired across Europe at the time of his death not just for his piety and diplomatic skills but also for his embrace of what would later become understood as scientific method. He wrote a number of remarkable treatises including *De Luce* on the metaphysics of light, *De lineis, angulis et figuris* on mathematical reasoning in the natural sciences, and *De irideis*, on optics and the rainbow. Grosseteste was frequently cited by Roger Bacon (1219–1292) a Franciscan friar whose work is based on empirical observation, citing Aristotle's emphasis on the collection of facts before deducting scientific truths.

Religion and the pursuit of knowledge and thought were not always comfortable bedfellows. Ibn Rushd (1126–1198) argued that matters of belief should be decided only on the basis of reason and evidence – a position deemed so dangerous that many of his books were banned by the religious authorities and he himself was briefly banished in 1195. But, overall, academic endeavour was located within the context of theological investigation both in the Muslim and Christian worlds during this period.

Crucially, the level of knowledge was insufficient to bring about fundamental changes in the human condition. When the Black Death swept across Europe, in 1348, there was very little understanding of the cause of the plague. Pope Clement IV, in Avignon, was advised by his physicians that it was the result of the conjunction of Saturn, Jupiter and Mars in 1341; others blamed the Jews, and others a miasma of bad air. The plague wiped out at least a third of Europe's population, and people were powerless to resist it, both then and in its many subsequent reappearances. Divine punishment was understood to be the proximate cause. Plaintive letters to the Pope from England in 1348 seek his emergency approval to the granting of absolution by lay people, as so many priests had died and the fear of going to hell as a result of dying unforgiven was real and abiding.

At this stage in history, humans are at the mercy of the elements. A series of poor harvests could easily lead to famine – indeed, the shortage of agricultural workers after the Black Death, unmilked cows in the fields, the wheat unharvested, was a cause of further deaths from starvation.

Perhaps the cataclysmic effects of the series of waves of plague gave greater urgency to the search for understanding of the world and the cosmos. But at this stage God was still in charge. At the end of the fourteenth century, in the Christian world, the Church remained the arbiter of acceptable investigation, while in the Muslim world a religious clampdown on intellectual activities seems to have brought the remarkable period of discovery and creativity to an end.

Icarus was, however, stretching his wings and learning to fly. The work of Copernicus (1473–1543), placing the sun at the centre of the universe with the earth going round it, attracted very little controversy, although both Luther and Calvin spoke against his theory. Galileo (1564–1642), following Copernicus, was attacked by Catholic theologians, but he was a pious Catholic himself, dedicating his polemic against his opponents *The Assayer* (1623) to Pope Urban VII, who received it with delight. His trial and condemnation for heresy, and his forced recantation, came later, in 1633, and was the result of a complex set of reasons both political and theological. But the symbolic significance of Galileo's condemnation resonated down the centuries, until the Vatican finally overturned it in 1992.

Francis Bacon (1561–1626) is widely seen as the 'Father of Experimental Philosophy'. His seminal work, *Novum Organum Scientiarum* (New Instrument of Science) was published in 1620. In it, he follows his predecessors such as Grosseteste and Roger Bacon by identifying observation and induction as the core of scientific method, and the use of artificial experiments as integral to the development of reliable conclusions. It is less well known that Bacon was a devout Anglican. In his *Essays* he wrote that 'a little philosophy inclineth man's mind to atheism, but depth in philosophy bringeth men's minds about to religion'. His death was caused by pneumonia, probably brought on when he was stuffing a chicken with snow while investigating the preservation of meat. Icarus falls.

Two more pieces in this jigsaw: first, the work of Leonardo Da Vinci: specifically, his drawing of the Vitruvian Man (1490): the classic expression of Protagoras' phrase 'man is the measure of all things': a foreshadowing of the Copernican revolution, placing humanity at the heart of philosophical endeavour, opening the door for a dethroning of God and the exaltation of the human.

Second, the work of Jean Bauhin (1541–1613) and his brother Gaspard Bauhin (1560–1624) followed and developed by Carl Linnaeus (1707–1778), who, in his work *Systema Naturae*, produced a comprehensive categorisation of the known world: a 'system of nature through the three kingdoms of nature, according to classes, orders, genera and species, with characters, differences, synonyms, places'.

Knowledge is power, as the saying goes. The exponential increase in knowledge between 1500 and 1800 – the period known with good reason as the Enlightenment – brought about a fundamental transformation in the relationship between humanity, the world and God. Icarus took flight; the fire of Prometheus was pressed into service to drive steam engines and blast furnaces; humankind discovered it had access to power beyond its most wild imaginings.

Financed by gold from the conquered people of the Americas, the newly opened trade routes to the East soon became congested with the ships of the European navies and the soldiers of European armies coming to conquer – powered by the rich seams of coal discovered beneath Europe's soil. Enlightenment was transformed into Industrial Revolution, and we discover, at the end of the eighteenth century and the beginning of the nineteenth, a triumphant merging of science and technology, pressed into the service of Progress. James Watt, credited with the invention of the steam engine, said he wanted to 'find out the weak side of nature and to vanquish her.' William Huskisson said in 1824, 'if the steam engine be the most powerful instrument in the hand of man to alter the face of the physical world, it operates at the same time as a powerful lever in forwarding the great cause of civilisation'. Huskisson was run over and fatally wounded by the steam engine *The Rocket*, at the opening of the Liverpool and Manchester Railway in 1830. Icarus flies, and Icarus falls.

We see, in the eighteenth and nineteenth centuries, a transformation in the relationship between humanity and nature. It is recognised by poets and philosophers, by scientists and inventors. It is an outworking of the Baconian creed that scientific method will provide the means of progress, of greater understanding of the God-given universe – developed by, among many others, Newton, who was able to show that the universe could be expressed through mathematical equations and was therefore knowable. The categorisation of nature developed by Linnaeus was a crucial step in a process of commodification of nature: from the location of mystery and the dominant context before which humanity had little control, the intellectual men (and they were, almost all, men) of Western Europe began to, as James Watt says, 'find out the weak side of nature and vanquish her'.

Walt Whitman, in *Songs of Parting*, writes

His daring foot is on land and sea everywhere,
he colonises the Pacific, the archipelagos,
With the steamship, the electric telegraph,
the newspaper, the wholesale engines of war,
With these and the world-spreading factories he interlinks all geography, all lands.

It is not by coincidence that the Romantic movement reaches its peak in the early nineteenth century: the work of Goethe (1749–1832) and

Wordsworth (1770–1850) was consciously or unconsciously the expression of the desire to maintain the natural world as a place where the sublime could be encountered. But perhaps the most prophetic counterpoint to the work of the Romantic poets was Mary Shelley's ground-breaking novel *Frankenstein or, The Modern Prometheus*, published in 1818. Frankenstein, the product of scientific endeavour gone horribly wrong, ending with the powerful image of the creature stumbling alone to his death through the mist and cold of the Arctic Circle.

But it is important to be clear that the onward march of scientific endeavour, combining, as it did in the early nineteenth century, with astonishing technological innovation, was not intended to be a negative process. While the Enlightenment philosophers, thinkers and activists firmly believed in a universal human nature, they also simultaneously championed tolerance, diversity, reason and the encouragement of science and technology. But the Enlightenment's emphasis on the free and independent private individual changes the relationship between individual, community and nature, resulting in greater inequality and more eager extraction.

The result, intended or otherwise, was an 'othering' of nature – the relationship was transformed, so that whereas in previous centuries humanity had been at the mercy of the elements and of nature, by the middle of the nineteenth century nature was a commodity to be exploited. You make holes in the ground and oil comes out – and gas, and coal, and iron ore, and a myriad of elements and compounds which enable humanity to straddle the globe like a Colossus.

Icarus takes flight. But not without the blessing and the sanctification of the Christian God. In an influential article published in 1967 – *The Historical Roots of the Ecological Crisis* – the American historian of science and medieval history, Lynn White, placed responsibility for the forthcoming crisis – already clearly discernible in 1967 – fairly and squarely at the feet of Christianity:

> Christianity inherited from Judaism … a striking story of creation. By gradual stages a loving and all-powerful God had created light and darkness, the heavenly bodies, the earth and all its plants, animals, birds and fishes. Finally, God had created Adam and, as an afterthought, Eve to keep man from being lonely. Man named all the animals, thus establishing his dominance over them. God

planned all of this explicitly for man's benefit and rule: no item in the physical creation had any purpose save to serve man's purposes.

Or:

Then God said, 'Let us make humankind* in our image, according to our likeness; and let them have dominion over the fish of the sea, and over the birds of the air, and over the cattle, and over all the wild animals of the earth,* and over every creeping thing that creeps upon the earth.'
[27] So God created humankind* in his image,
in the image of God he created them;*
male and female he created them.
[28]God blessed them, and God said to them, 'Be fruitful and multiply, and fill the earth and subdue it;

The word 'dominion' is a translation of the Hebrew word *radah*, and 'subdue', of *kavash*. Scholars have challenged an overly simplistic interpretation of the translations – many would say that *radah* has overtones of stewardship, and *kavash* of care – but very few would deny that the sense of this crucial passage in the Christian and Jewish scriptures places humanity at the pinnacle of the hierarchy of creation, with absolute power. In the image of God he created them.

To be clear: I am not suggesting that scientific and intellectual progress is intrinsically bad. The transformation in life expectancy, health, standard of living, technological progress and the explosion of creativity – in the rich world, at least — these are good things for humanity. Remarkable innovations and inventions in the nineteenth and twentieth century have enabled humans to reach pinnacles of achievement unimaginable to our forebears. Now, we tend to take them for granted, when thousands of people climb Mount Everest every year and astronauts fly routinely into space and usually, return safe. But the sense of achievement, of mastery over nature when Edmund Hillary and Sherpa Tensing reached the summit of Everest on 29 May 1953 and stood on top of the world, (or rather, when the news reached the rest of the world on 2 June, the day of the Queen's coronation) was equalled when, on 20 July 1969, Neil Armstrong and Buzz Aldrin flew closer to the sun than anyone ever had before, and took the first steps on the moon.

There were straws in the wind which indicated that all was not as good as it could be. For example, London became a byword for impenetrable fogs in the nineteen and early twentieth centuries. Regular pea-soupers, the consequence of millions of coal fires spewing out smoke which became trapped in the bowl of the Thames basin – culminating in the Great Smog of 1952 – were the causes of extreme distress, respiratory disease and death. William Blake's 'dark Satanic mills' that he describes in *Jerusalem* were not, as is often assumed, in the industrial north – they surrounded his home in Lambeth. The consequences of industrialisation, especially in terms of forced labour, child exploitation, and urban poverty, caused much concern in the nineteenth century: but underlying the concern and the anxiety was a profound belief that better policies and well-formed reforms of abuses would solve the problems. The Clean Air Act of 1956 made pea-soupers a thing of the past: the work of philanthropists outlawed the employment of children, at least in Britain if not in the Empire: and the gradual development of the welfare state was given a huge boost after the end of the second World War, with the introduction of universal health care and free education, in order to help create 'a land fit for heroes to live in.'

At the heart of the belief in progress was – is – the belief that nature can be tamed. The significance of the conquering of Everest and the Moon landings is that they demonstrated that, for humanity, there are apparently no limits in our ability to realise our often unimagined potential. They serve the metaphor of triumphant progress, and the rare events which seem to reassert the power of nature over technology – the sinking of the Titanic or the explosion of Challenger in 1986 – are soon forgotten, buried under an onslaught of news of successful new inventions and innovations.

My own recollection bears this out. As a schoolboy, in the 1970s, I remember very clearly discussing the apocalyptic threat of the time – the risk of running out of oil, which would, we feared, bring the potential for human progress to an end. I also remember very clearly having conversations about pollution, of the atmosphere and of the sea – in fact, I took part in a nationwide survey of lichen, to assess the incidence of pollution across the country. Pollution was taken seriously; but underlying the concerns was a clear sense, amongst those who taught me and those with whom I discussed such things, that nothing we could do would fundamentally harm the atmosphere – and, even more crucially, that the sea was so big that nothing we could do could create any lasting damage. The world, we thought, had a self correcting mechanism: Gaia theory, emerging in the 1970s, seemed to bear out the idea that the environment we were actively taming would, in the long run, correct any damage and remain healthy.

Icarus was flying. Icarus flies still, high up in the outer atmosphere, joined by squadrons of Boeings and Airbus and swarms of satellites. He spreads his wings and, to change the metaphor, puts a girdle round the earth in forty minutes. We are now Prospero the great magician: we have tamed Puck and subdued Caliban; we have captured the island. We are God. We have asserted the dominion which is our divine right. We have multiplied, we have filled the earth and we have subdued it.

In an influential and farsighted book published in 1984, the philosopher and ethicist Hans Jonas recognised that humanity was in imminent danger of flying to close to the sun. Modern technology, he said, has 'opened up a whole new dimension of ethical relevance for which there is no precedence in the standards and canons of traditional ethics'. Traditional ethics, he argued, has always worked on a horizontal time frame – a time frame of immediacy:

> 'Love thy neighbour as thyself': 'Do unto others as you would wish them to do unto you': 'Never treat your fellow man [sic] as a means only but always also as an end in himself' – and so on. Note that in all these maxims the agent and the 'other' of his action are *sharers of a common present*.

Ethics have always been understood to be immediately reciprocal: insofar as ethical actions imply a general perception of the common good, even if this common good is understood to have universal characteristics, that good is understood always to be translated into the here and now:

> Precisely because the human good, known in its generality, is the same for all time, its realisation or violation takes place at each time, and its complete locus is always the present.

> All this has decisively changed. Modern technology has introduced actions of such novel scale, objects and consequences that the framework of former ethics can no longer contain them.

Jonas argues that we have to develop a new sense of responsibility – not only to those who are living in the same time-frame as us but, crucially, also to those who are as yet unborn. We have to consider the long-term consequences of our actions, in terms of generations, not of years: we have, he argues, a responsibility to the humankind of the future – indeed, not just to humankind but also to the biosphere.

Ethically, it is not simple to argue that we have a responsibility to the as yet unborn, for they do not yet exist. Jonas argues, building on the Kantian principle of the categorical imperative in his admittedly dense book, that the existence of life implies an ethical responsibility to ensure its continuation and its flourishing: that we, as humans, are part of the cycle of life, and that we should have, as an overriding ethical principle, the requirement to act 'so that the effects of our action are compatible with the permanence of genuine human life'.

It is, he argues, this crucial element of ethical responsibility – to the future, to those as yet unborn – which we have failed to take into account in the actions which we have, as a race, undertaken since science and technology combined and unleashed unheard of capacity upon the earth. Jonas was writing in the 1980s when the most immediate manifestation of that was – as it remains – the potential for total nuclear holocaust: now, forty years later, we can definitively add the potential for environmental catastrophe to the potential for nuclear destruction.

IG Simmons, environmental historian and geographer, argues that free market economics have contributed to the current situation:

> The neo-classical economic theory developed after Adam Smith is distinguished (among other ways) by its absence of meaning taken from other spheres of life. It becomes a closed and self-referential web of concepts that dispelled morality from human livelihood ... It thus placed no obstacles in the way of accumulation of wealth.

This essay takes a 'grand narrative' approach to human history. I would argue that this approach is justified in the context of the ecological crisis which faces us, but it is not a theoretical approach which commands universal assent. Almost everyone I have cited has been male. Perhaps one of the explanations for our current predicament has been the archetypally masculine linkage between knowledge and power, expressed so forcefully by William Huskisson ten years before he was run over and killed by a steam train.

The connection between immediate action and immediate returns and the ethical failure to consider future generations is alive and well, manifested most clearly by the environmental hooliganism of the Trump administration, but equally by the all too frequent failure of all of us, as individuals to consider the long term consequences of our lifestyle and choices.

We are all Icarus now: collectively and individually, we are approaching the sun and the wax is melting on our wings. We have become disciples of Prometheus, harnessing fire to our own ends, and have failed to comprehend the consequences of our decisions.

Is it too late? Have we lost too many feathers from our wings: are we about to stall and fall headlong into the ocean? The Pope, in a recent speech in the Vatican, said this:

> The United Nations ... proposes integrating all the [Sustainable Development] goals through the 'five Ps' people, planet, prosperity, peace and partnership... I welcome this approach, which can also help to save us from an understanding of prosperity that is based on the myth of unlimited growth and consumption ... There are still people who stubbornly uphold this myth, and who tell us that social and ecological problems will solve themselves simply by the application of new technologies. An integral approach tells us that this is not true.

The Pope calls for an 'ecological conversion' in his classic encyclical, Laudato Si, published just before the Paris climate change talks in 2015. Published the same year, the Islamic Declaration on Climate Change concludes:

> We call on all Muslims wherever they may be –
>
> Heads of state
> Political leaders
> Business community
> UNFCCC delegates
> Religious leaders and scholars
> Mosque congregations
> Islamic endowments (*awqāf*)
> Educators and educational institutions
> Community leaders
> Civil society activists
> Non-governmental organisations
> Communicators and media

> to tackle habits, mindsets, and the root causes of climate change, environmental degradation, and the loss of biodiversity in their particular spheres of influence, following the example of the Prophet Muhammad (peace and blessings be upon him), and bring about a resolution to the challenges that now face us.

We hear from global religious leaders and we hear from children and young people: this is our future, they say, and we want to flourish in it. We hear calls for the re-enchantment of nature, and passionate arguments for the re-wilding of over-engineered and exploited landscapes. There is renewed interested in the great prophets of care for creation – St Francis, Teilhard de Chardin, Gandhi.

On the ground, there are signs of hope. I chair the Faith for the Climate Network, created in 2014 to encourage closer cooperation on climate change between faith traditions. The network includes Christian, Hindu, Muslim, Buddhist, Sikh and Jewish faith groups. It has, through the provision of better resources, symposia and joint actions such as the Pilgrimage to Paris before the Climate Change Conference in 2015, begun to make a material difference to the response of faith groups to the challenges we face. The Church of England has helped to develop an investment tool, the Transition Pathway Initiative (TPI), which is intended to transform its investments, moving towards a low carbon economy. The TPI covers US$3 trillion worth of investments across the world.

These are all, in different ways, recognition that we are flying too high. They are straws in the wind, hints of hope that we may, in the end, be more like Daedalus than Icarus – that we may find the capacity to limit our ambitions and fly neither too near the sea nor too close to the sun. The all-too-real prospect of environmental meltdown poses a whole new set of philosophical questions, but underlying them all is the challenge to the ever increasing population of humans on this planet: whether we have the capacity to transform our consumption for the sake of the generations after us, and whether we can listen to the story of Icarus and amend our ways before it is too late.

GAIA 2

Christopher B Jones

Global climate change is a trending topic, with atmospheric carbon dioxide rising to 415 parts per million in the atmosphere, growing human impacts from sea level rise and atmospheric warming, and dire warnings from scientific and international panels that action needs to be taken soon to avert global ecological disaster. It is in this context that ideas emerge: the notion that ever weirder weather is to be expected, that action is required on a planetary scale to ensure humanity's survival, and the assertion that our species has set forces in motion that will forever change individual and collective human behaviour.

The concept of 'global weirding' has grown in popularity as an alternative to the phrase 'global warming'. The term was first attributed to American renewable energy guru, Avery Lovins, and then popularised by the centre-right economist Thomas Friedman in an op-ed piece for the *New York Times*. The neologism appears to have great currency given the incipient effects of global climate change, and the examples of weather, sea level rise, glacial melting and other events that exceed the norm. Examples include the highly variable atmospheric jet stream, polar vortex in the northern hemisphere, and other extremes in temperature, rainfall, wildfires, and flooding across the planet. Therefore, global weirding was intended to broaden the debate beyond sea level rise and abstract discourse about warming average temperatures and growing greenhouse gas emissions. David Wallace-Wells in 2019 catalogued the increasing costs of higher average temperatures and or feedback loops (potential black swan events) that could exacerbate warming even further. Among climate researchers and advocates, there appears to be consensus that global weirding and what Wallace-Wells calls 'cascading catastrophes' will occur if we do not act quickly, and are perhaps even likely given what we have already set in motion. The next question becomes what do we do about it, and the answer is complicated. The example of potential feedback loops to accelerate the warming process are considerable, including dramatic shifts in the Earth's albedo (or reflectiveness), methane release from permafrost and warming oceans, undoubtedly weird events if and when they happen.

Global weirding, as it has been addressed in the literature, has mostly referred to climate weirding, that is, the physical impacts of climate. But this weirding also extends to how we respond to climate change, particularly nascent attempts to affect geo-engineering and the growing international interest in approaches to technological fixes. We can even extend global weirding to consider the extinction connections to human-induced climate change, such as shifting climate zones, earlier spring, and longer fire seasons, movement of invasive species, pathogens, insects, and animals outside of their normal habitats. American futurist and postnormal times theorist John Sweeny expanded the notion of global weirding further by describing it as a 'meshwork' that includes not only cascading catastrophes, but also a growing technosphere upon which we are increasingly dependent, and the growing transnational drive and reach of postnormal actors.

To put global weirding in context we need to examine climate change over historic geological periods, and the best place to begin is the Gaia theory proposed by the British scientist and environmentalist, James Lovelock. We need to explore postnormal times (PNT) theory as PNT appears to me to be both an artefact of global weirding and a driver of it, because complexity, chaos, and contradiction are both features of the postnormal times we are living in, and will also have an impact on how we respond to climate change and our behaviour in a warming world. PNT theory, particularly postnormal creep, and the postnormal menagerie (black swans, jellyfish, and elephants) are useful tools to understand the challenges ahead of us and to better navigate our way to preferred futures.

According to Lovelock, the Earth's biosphere, atmosphere, and geologic systems are an integrated cybernetic system, with multiple, complex feedback processes. Some of those processes, such as glacial periodicity, deep ocean current circulation, carbon deposition in the oceans and limestone, plus many others, have time frames that extend from tens of thousands to hundreds of thousands of years in length. Key to Gaia theory is our understanding from cosmology of the evolution of our sun.

According to cosmology, the sun has increased in overall solar radiation by 30% since the beginning of life on the planet. While solar radiation received by the planet has increased, the overall temperature on earth has neither been cold enough to freeze the oceans entirely, or too hot for life to exist, nominally the boiling point of water (although some exotic microbes do manage). In other words, the greenhouse effect was an important regulatory process in early Earth history, but is of less value as the sun heats up.

Lovelock demonstrated how Gaia is not teleological or purposeful, but responds to sometimes fairly simple rules. The best example of this was his Daisy World model that showed how even two different colours (black and white) of daisies could regulate planetary temperature. The point is that while we may not, as a species, grasp all of the fine points and subtle nuances of the global chemical and biological systems, the broad parameters and dynamics can be understood, and arguably must be better understood collectively for us to survive as a species.

Gaia has provided a complex, billion-year-old system of feedback loops between biota and chemical cycles. This robust system has protected life as a whole from extremely violent and endogenous and exogenous forces. Endogenous planetary events have included: one of the first extinctions caused by the production of oxygen in the atmosphere, super volcanism, and changes in ocean currents due to plate tectonics. Outside exogenous events have included impacts with comets and asteroids, variations in the sun's radiation, and cosmic dust storms. While the geological record shows vast periods of volcanism, resulting in periodic mass extinctions, yet there has been a continuation of life and greater diversity and complexity of life forms. Lovelock argued that our species' experiment with the atmosphere may propel the planet into a higher state of thermodynamic equilibrium. He argued that we should remember lessons from the planet's past. In his last book, *A Rough Guide to the Future*, Lovelock urged us to remember the past, particularly the Earth's climate 55 million years ago during the Paleocene-Eocene Thermal Maximum (PETM) when the planet was 6 to 8°C more than it is today. He described the planet then as a tropical system with no ice at the poles, with a highly coupled atmospheric system of giant Hadley cells that were like thermal pipelines between the equator and the poles, distributing solar energy. It is an ominous possible scenario for climate weirding in the not-so-distant future.

The Gaia theory argues that our species plays a new role in the climate system. For at least one million years our progenitors had little impact on the climate, and *homo sapiens* and archaic human species' migration patterns responded to the glacial cycles and sea level changes. Macro historians have described the cycles, rhythms, and evolution of human societies, yet agriculture and civilisation only emerged after the end of the last glacial cycle. In other words, the rise and fall of previous civilisations occurred during relatively moderate interglacial climate. Macro history and Gaia are useful to provide us with Big Picture contexts and underscore the importance of long timescales and broad horizons to consider the survival of modern industrial civilisation, or even our entire species. But the impacts of human migration across the planet, first agriculture, and then industrial activities and resource extraction from the planet have altered its face. We are advised to consider those long-term timescales and how climate change is a window into human values that are poorly aligned with planetary systems.

Also relevant to the futures of climate weirding is the famous 1972 report, *The Limits to Growth.* Funded by the Volkswagen Foundation and commissioned by the Club of Rome, it was the work of a Massachusetts Institute of Technology (MIT) team, headed by Donella H Meadows. The MIT scientists used a computer simulation model, dubbed World 3, to study the consequences of interaction between the earth and human systems focusing on population, pollution, food production, industrialisation and consumption of natural resources. Much-maligned and savagely attacked by conservatives and industrialists, the report of the MIT team posited that normal *standard runs* of the World 3 model forecast that population growth and industrial development would exceed carrying capacity by the mid-twenty first century. Lost in the controversy was the assumption of the modelers and futurists addressing the *global problematique* (species extinction, desertification, deforestation, pollution, climate change) that they were not *predicting* the future, but exploring the parameters of and consequences of system change. A member of the team, Joel Barker, wrote about his role in the MIT study as a young researcher. He noted that there was a lot of attention paid to assumptions and weaknesses of the models, but more importantly it is the lesson from the similarities between the models and realities, that complex phenomenon are full of delays and time lag, which 'blind us to large-scale change'. There are cumulative effects of billions more humans on the planet, increased patterns of consumption, resource use, and energy in the system. Those additional humans and complexities will create further uncertainty.

Climate has to be seen in a range of temporal time frames, from long-term glaciation (due to plate tectonics), 100,000 year glacial periods interrupted by warmer interglacial periods lasting on average 10 to 12,000 years, and shorter-term changes of solar radiation and volcanism (likely the cause of a mini-ice age in the sixth century). Thus anthropomorphic changes have to be seen in a broad context. The scientific consensus is that we would nominally be entering a new glacial period and ending the warmer interglacial period. The more recent periodicity of glacial cycles over the last million years appears to be causally linked to the nineteenth century Serbian astronomer Milutin Milankovitch's observation of cycles of axial tilt, eccentricity, and precession that force cooling by taking the northern hemisphere the furthest away in its orientation to the sun over a roughly 100,000 year period. There appears to be a general consensus that over the last million years, the generally stable homeostasis has cycled between glacial cycles and short interglacial cycles, with the Milankovitch effect triggering each new cold phase. Lovelock has argued that we may be fundamentally altering this pattern. Climate weirding becomes geomorphic weirding if humans alter normal glacial cycles.

The astrophysical and geological records support this theory and ironically, perhaps a contradiction, is that according to ice core analysis the Earth's atmosphere appears to have had higher concentrations of carbon dioxide immediately preceding glacier building in the northern hemisphere at the beginning of each of these cycles. One attempt to explain terrestrial feedback loops and ice ages was the work of John Hamaker, American engineer and climatologist, who argued that planetary biomass in glacial times is more productive than interglacial periods, due to the abundance of ocean life, particularly diatoms and other microscopic life that thrive on dust blown from continents. The key variable is ground rock dust, a product of glacial building and retreat, enormous dust storms produced between the warm tropics and increasingly cold polar north.

According to Hamaker, during interglacial times the ground up rock dust is blown across the lower temperate zones and accumulates, creating rich soils such as the loess soils of China that become depleted of minerals over the 10 to 12,000 years of the interglacial period. As demineralisation leaves temperate and alpine forests less healthy, insects and fire take their toll. Hamaker believed that the normal end of the interglacial would be characterised by massive forest fires that would actually raise the level of carbon dioxide. Massive fires would affect the solar albedo, and he argued that the reflection solar radiation from these massive firestorm clouds would actually trigger global cooling in alignment with the orbital forcing of Milankovitch cycles. Hamaker predicted a return to glaciation by the end of last century, and in spite of ever increasing volumes of smoke – particulate matter – in the atmosphere, global warming accelerates. His theory is perhaps a reminder that we need to recognise our limitations in understanding the various positive and negative feedback loops, and consider the outlier possibilities. We would do well to remember that climate history reveals that return to glaciation can happen quickly. This may be an example of the unknown unknowns that face us in a weirding world.

Another parallel geophysical actor in the climate picture is the deep ocean current. The deep ocean current story extends back hundreds of millions of years before continental drift remade the face of planet Earth. Conventional scientific consensus is that prior to the breakup of Pangaea – the supercontinent that existed during the late Paleozoic and early Mesozoic eras – the Earth's climate was much more tropical, but there were also periods of much colder climate. Continental drift and the emergence of a deep ocean current appear to have had a moderating effect on the overall planetary temperature. Without the deep ocean current, there would not be local currents at the ocean surface, such as the Gulf Stream, that distribute heat across the North Atlantic. Similar currents operate in other oceans, as well. Scientists believe that the absence of that deep ocean current has been the source of rapid planetary cooling in the geological record. For example, at the end of the last glacial period, ice dams collapsed in northern Canada that released vast amounts of fresh water into the North Atlantic that appear to have stalled the deep ocean current, creating a mini ice age as the planet was entering the interglacial period.

Our understanding of the geological and climate history has improved recently thanks to ice core sampling, ocean floor drilling, and tree ring and other dating processes. There are, however, lots of ifs and buts - obviously a lot we still do not know, and this ignorance can be compounded with all the uncertainties involved. Yet, we are now engaged in the grandest laboratory experiment ever, using the entire atmosphere as a laboratory. There is little doubt that humans have become a species with global impact, but given the complexities, the uncertainties, and the contradictions in our behaviour, we cannot know what the consequences will be. Is our experiment only forestalling or inhibiting or even accelerating the return of northern hemisphere glaciers? Or will the experiment result in accelerating warming, perhaps reestablishing the climate regime during the PETM 55 million years ago?

We have yet to create, generate, or adopt a collective mythology that is consistent with Gaia theory. For the past decade or so, the trending neologism to describe the coming limits to growth and what researchers describes as the coming 'biological annihilation', or the Sixth Mass Extinction, has been the word Anthropocene. The word acknowledges the planetary impact and long-term consequences of human civilisation. American feminist and science critic, Donna Haraway, suggests that Chthulucene is a better neologism to replace Anthropocene as the proper term to identify the geological period to follow the Holocene. Chthulucene, from the Greek *chthonios* means: 'of, in, or under the earth and the seas'. 'The chthonic ones', Horroway notes, 'are precisely not sky gods, not a foundation for the Olympiad, not friends to the Anthropocene.' Haraway takes the radical view that we need to embrace our grief and face the realities of the interconnected, intertwined, web of life that includes other species, fish, animals, birds, insects, and other living parts of the biosphere as kin. We need to appreciate not just that we have become the dominant predatory species on the planet, but that we need to put our hugely successful skills at adaptation and resilience to work in service of the planet instead of seeing it as a resource. Thus, we are not only entangled with the climate, but also in webs of life that dynamically change and evolve. We have already set the sixth extinction in motion and will have to deal with the consequences of that over millennia. More time lag. Kurt Vonnegut would say 'and so it goes'.

We can contribute to better regulating of feedback in the Gaia system, and potentially make it smarter. We will have to find the wisdom to get there.

The most recent report of the United Nations and the Intergovernmental Panel on Climate Change (IPCC) made dire predictions for the planetary climate by 2050 if zero omissions of carbon dioxide are not realised by 2030. Sadly, it appears the reduction of carbon dioxide production to preindustrial levels is not only unlikely, but there has been an increase in the overall production of greenhouse gases and their release into the atmosphere. It is argued that the 2050 timeline closely aligns with the average run of the MIT World 3 model, as noted earlier, and it begs the question of whether the Earth's carrying capacity has actually already been passed. Mark Lynas, the British author and journalist, provides a litany of worsening environmental consequences for each additional degree of warming past the preindustrial baseline. Even the smallest of changes, even 2°C to 3°C – not far from the nominal projections of the IPCC – will have enormous negative consequences. Heating more than 6°C above pre-industrial levels would result in civilisational catastrophe, and yet the lag in carbon dioxide absorption in the environment means that even if we achieved zero emissions tomorrow, it would take decades, if not centuries for carbon dioxide levels to stabilise. And we are not even accounting for the other major greenhouse gases such as water vapour, methane, chlorofluorocarbons that continue to accelerate warming. Those are examples of the potential reservoirs of uncertainty in the current climate models.

As I write, there are extreme flooding events throughout the Mississippi and Missouri River basins in the US, widespread flooding in East Central Africa, and extreme heat and drought in China and Australia. Examples of extreme weather events now abound in a litany of multiple destructive cyclones and hurricanes, drought, wildfires, and even increasing tornado activity. Climate catastrophe is happening now! It is not something out there in 'the future'. Part of the reason I am drawn to the Gaia theory is that it provides a scientific basis for the claims of accelerating warming, a measurable cybernetic system, that follows fairly simple rules of systems, although there are layers of complexity, nested biological, chemical, and thermal systems that in concert sustain a larger dynamic, and somewhat stable system – over the millennia.

We can see some of those systems unravelling. As the oceans become more acidic, the result will be less diversity as ocean reef communities die, potentially releasing undersea continental methane ice, releasing methane from the tundra and Arctic areas, and of course the loss of trees from Australasia and South America not only release more carbon dioxide but are also lost as a carbon sink. It results in a runaway train phenomenon, where the warming creates a vicious negative feedback loop, and it tends to become self-reinforcing. Lovelock posits that there are mechanisms that may eventually check those feedback processes, but it could take centuries, if not millennia, to return to stability, even if we were able to reverse or halt accelerated warming.

And so to postnormal times (PNT) – the period of accelerating change and disorder we find ourselves in. PNT theory holds that growing complexity, chaos, and contradiction inherently create greater uncertainty about scientific and policy decisions. Furthermore, a postnormal creep tends to shift the baseline, so that increasingly the unexpected happens. That clearly fits the model of global weirding and accelerating climate change. Some of the models for the tools of PNT theory, such as black swan, black jellyfish, and black elephant are drawn from climate change or apply to it.

The black elephant (in the room) metaphor is a classic point in case, particularly in the political culture of the United States of America, where the captains of industry and political leadership currently reject the idea of anthropogenic climate change. The elephant in the room is not only climate change itself, but the use of fossil fuels, consumer culture, and even fundamental economic models as well. The idea of growth itself is being questioned by social critics and the environmental movement. Measures of economic growth, for example, are being questioned because they account for the economic costs of recovery and disaster. Disaster should not be good for business. David Wallace-Wells noted the analysis of the actual economic costs of climate change by the end of the century are likely to exceed the total wealth of civilisation.

From a radical deep ecology perspective, it has even been argued that not only is civilisation unsustainable, but even agriculture may be at the root of our problem as a species. There are ample social critics of progress and civilisation that have fuelled that fire, such as Diamond, 2005; Ehrlich & Ehrlich, 2013; Slaughter, 2010; Wallace-Wells, 2019; Wright, 2004.

There are those that advocate for return to a Paleolithic, hunter-gatherer lifestyle. It is an extreme view, but suggests that even civilisation, as a concept, may be an elephant in the room. Another elephant in the room, in the shared space of climate and culture may be the negative images and apocalyptic cultural memes that drive popular culture. For example, in the USA, there is a sizable population that holds that End Times are coming. Such beliefs could accelerate careless behaviour and encourage a 'party till the end' mentality.

Many coming changes will be of the black jellyfish sort, unexpected high impact events or developments driven by positive feedback loops to produce even further chaos. Examples of that include melting permafrost that have had serious consequences for roads and bridges, the potential for melting ice and permafrost to release pathogens for which humans have few immunities, and the multiplier effect of methane released as permafrost melts. In recent decades, the permafrost zone has moved hundreds of miles further north towards the poles.

Invasive species have been spread around the globe thanks to global supply chains and transportation networks. Climate change only exacerbates that problem as tropical and subtropical species move further north with warmer temperatures. Forecasts for the 2019 summer in the Great Lakes anticipate large algae blooms due to the wet spring and flushing of phosphates and agricultural runoff into the lakes. Florida and the United States' Gulf Coast continue to struggle with algae blooms, while coral bleaching continues across wide swathes of the oceans. Instances of black jellyfish events will become the norm in a postnormal global weirding environment.

Black swan events, outlier events and developments appear to arrive 'out of the blue,' with the potential for disruptive impacts at a systemic level. The near total destruction of the water and electrical infrastructure of Puerto Rico after two hurricanes is one example. A sudden shift in the Earth's magnetic field has been suggested as one such possible black swan planetary-scale event. The collapse of the deep ocean current might be another. Bee colony collapse disorder, and the disappearance of pollinators, may be harbingers of more systemic ecosystem failures.

One of the key messages of postnormal times theory, particularly postnormal creep – increasing numbers of unexpected black swan and black jellyfish events – is that the breadth and scope of uncertainty will grow. Events and discoveries continue to unfold showing our limited understanding and outright ignorance about Gaian systems. One recent discovery by glaciologist Michael Willis found that even glaciers in cold, dry areas are being affected by global warming. Since 2013, the outlet glacier on the Vavilov Ice Cap has shifted from advancing 60 feet a year to 60 feet per day. 'The fact that an apparently stable, cold-based glacier suddenly went from moving 20 metres per year to 20 metres per day was extremely unusual, perhaps unprecedented,' Willis said. 'The numbers here are simply nuts. Before this happened, as far as I knew, cold-based glaciers simply didn't do that ... couldn't do that.' This is just one example of the litany of recent discoveries about the acceleration and pace of melting of Alpine glaciers, Greenland ice sheets, both West and East Antarctic glaciers that consistently exceed previous expectations.

Postnormal times may parallel longer-scale planetary changes. For example, there may be a kind of punctuated equilibrium as humans and planet co-evolve and seek resilience during the coming troubles. Haraway's advice is that we 'stay with the trouble' and come to terms with the fact that a new era, the Chthulucene, is underway. There does appear to be growing consensus that the midcentury is likely to present this culmination of forces, playing out of variables, a hyper postnormal times period, beyond which we will either see our civilisation decline and collapse, or be transformed by some Singularity beyond our current imagination, or transformed to a Green Gaian society. All are still possible at this point. However, increased complexity and unknown unknowns, more black swans, may emerge. Those may preclude other preferred futures we might desire now, and may regret foreclosing on in the distant future.

Gaia is an extremely complex system, that has over time balanced fairly fine-tuned geochemical and biological processes that now sustain life on earth by balancing temperature, atmospheric chemistry, ocean chemistry, and cycling minerals through the Earth's crust. There are vast changes that have occurred due to the different species that have evolved over the planet's history, but the basic patterns are mostly driven by microorganisms, another key element in planetary functioning of which most people have very little knowledge or awareness. The forces of nature are also obviously very chaotic, as can be seen in rock deformation and the effects of plate tectonics deep within the Earth's crust. The forces of contradiction, between neither too hot nor too cold over a billion years have managed to find some balance. We need to keep these elements constantly in mind. But we need wisdom, not just know how.

As the dominant species on the planet, as a species that has managed to live in outer space, harness the atom, and now alter our own genome, we need to figure out a way to live in harmony with the planet and in a way that does not destroy the regulatory system that keeps the glacial cycle working. We were blessed to have civilisation emerged during a benign interglacial period, but we may need to come to terms with the fact that glaciation is what is normal. We truly have been living in postnormal times during interglacial times. Glacial is normal. If that were true, what would it mean? What would our responsibility be as a species if, to sustain life on the planet, we are obligated to keep it cooler?

Lovelock laid out many of his concerns about the consequences of our inability to rein in carbon dioxide emissions and made a strong argument that IPCC forecasts were far too conservative. He posited that positive feedback loops, such as methane release from permafrost, changing albedo, and other large-scale physical processes were already accelerating, well underway, and have self reinforcing tendencies that could potentially take humanity in short order to a much warmer thermal regime, potentially as much as 6–8° C warmer. That would likely leave Antarctica as the only habitable place left on Earth. Terraforming Mars might then be more than just science fiction, but a good investment. The point is that Gaia theory suggests that Earth may go on just fine without us, that is the biosphere, the biomass as measured by the totality of creatures in the ocean and on land will probably still exist, it just would not be hospitable for humans.

I hope Lovelock is wrong; and it is not too late. I also am optimistic that we can find a way to mitigate the damage we have caused, yet it will take a global effort, a paradigm shift of planetary proportions, a true global brain in some form of scientific Gaia 2.0. Meantime, the global climate weirding will continue to add to the acceleration, uncertainty, chaos, complexity, and contradiction of postnormal times. Welcome to the Chthulucene, all of our Gaian kin.

IN LOVE AND RAGE

James Brooks

1.

On 17 November 2018, my life changed.

Until then, for most of it, I'd felt like a spy abandoned behind enemy lines. The organisation I worked for had long collapsed and my contacts had vanished but I was still out there, living in a world that didn't apply to me, from which I was exempt, still soaking up the information, waiting for the call when I could tell all and be understood.

My biggest problem was that while this organisation was dead, its ideology was still programmed into my brain, dictating how I saw the world.

You're programmed too, by the way, but your programming is better aligned to the external operating system and you don't notice.

For example, unlike you, I don't have a smartphone.

'Why not?' you say when you notice this.

'Because I don't want one.'

'Why don't you want one?'

'Why don't you want an emu on a piece of string?' I want to answer but don't. 'Same question.'

A lot of it, I feel, has to do with wanting things, or in my case, not wanting them. I don't want a smartphone; I don't want new clothes; I don't want a watch that tracks my location and communicates with my central heating to turn it on when I'm five minutes away from my flat.

I've read about the French social theorist René Girard and his theory of mimetic desire; everybody else wants these things so I should want them too. But I don't. Why not? Because of the programming.

You probably think that I'm a deeply weird person by now and I don't blame you, but what you've got to remember is that I was a spy for twenty years, and I've avoided detection. I look and act very much as someone like me – a white, middle-aged, middle-class man who lives in the London suburbs – should do. There are undoubtedly quirks, like not having a smartphone, but nothing outrageous. I'm normal. You wouldn't spot me.

But on 17 November 2018, my life changed. I learnt that my organisation wasn't dead. I began my journey back in from the cold.

It had been billed as 'Rebellion Day' by the organisers, a group called Extinction Rebellion that no-one I knew had heard of. Now, after blocking off roads in four major locations in London for over a week, racking up more than 1,000 arrests in the process, lots of people have heard of them and call them, rather snazzily, 'XR'. Back then, though, it was the full 'Extinction Rebellion'– six or seven ill-fitted syllables making quite a mouthful.

At 9.45 that Saturday morning I walked onto Westminster Bridge from the southern side, just up from Waterloo station. It was quiet; a fine, chilly but clear November morning. I stepped past a couple of policemen in yellow vests who, happily, didn't register me. I walked across the bridge quite normally, as if I wanted to get to the other side, with the three-point instructions I'd read on Extinction Rebellion's 'Rebellion Day' Facebook page playing in my head.

Now on the bridge, I was up to point two: 'Wander innocently up and down the bridge on the pedestrian walkways. Take selfies, admire the view. Don't congregate in large groups. Wait for the signal…'

I'd come across Extinction Rebellion – how else? – on the Internet. I think it was a Tweet from a Quaker account that first tipped me off. I wasn't a Quaker but I'd gone to a Quaker secondary school and had an abiding fondness for this tiny Christian movement best known for its unyielding pacifism. Every couple of years, my fondness would bloom into an appearance at a Quaker meeting where some Quaker would stand up and talk about 'God' and I'd decide this really wasn't for me and not come back. Twitter was safe, though.

I tell people now: 'I'd always been concerned about the environment', but that isn't quite right. I'd always despaired about the environment. As I saw it, mankind had unleashed this great destructive machine over the Earth, gobbling up forests and animals and sea-life, pumping it all back out as toxic clouds of black smoke, that no-one wanted to stop. This was the machine that made aeroplanes and crisp packets; tennis balls and tins of tuna; lollipops and laptop computers and everything else that we live on, everything that spells progress. What could I – one in seven-and-a-half billion – do about that?

Even the supposed environmental movement was fuelling the great machine. I remember a music festival where Greenpeace had a stand. They had some fancy set-up with virtual reality headsets and a mini-cinema and you could buy t-shirts and get a tote bag. That kind of needless consumption was OK, yeah?

What wasn't part of the great machine? Extinction Rebellion, apparently. I couldn't believe it at first. I headed to their website expecting to see logos of NGOs or charities planted in the navigation bar at the bottom, but there were none.

At the centre of Extinction Rebellion, I discovered, was Roger Hallam, a PhD student at King's College London and in his early fifties. He'd been researching what kind of protest movement would have the greatest chance of success against the odds. The answer, he discovered, lay in mass-participation and non-violent civil disobedience, with each of the two being equally critical. *Mass participation*: 3.5 per cent of the population needs to join up (this statistic is actually from historical research by Harvard political scientist Erica Chenoweth). *Non-violent civil disobedience*: disruption, mass arrests and incarcerations are essential, for without them power will not budge, but participants must stay entirely peaceful or risk losing wider public support.

Together with a rag-tag bunch of activists, Hallam was putting his research into practice to achieve three goals. First: get the UK government to tell the truth on the state of the climate and ecological emergency. Second: commit the UK to carbon-neutrality (that is, the UK would produce no net carbon dioxide) by 2025 and halt biodiversity decline. Third: the transition to a carbon-neutral economy should not be overseen by the government but instead by citizens' councils – randomly selected people from all over the country meeting to decide how to proceed.

Clicking about online, I found an article Hallam had written for *The Guardian* back in June 2018, berating the government over its decision to build a third runway at Heathrow airport . 'CO_2 emissions are altering our planet,' Hallam wrote, 'and it will lead to humanity's destruction unless we do something about it'. Limiting global warming to below two degrees above pre-industrial levels is a central goal of the 2015 United Nations Paris Agreement on climate change, but look at the science, said Hallam – two degrees of global warming is already locked into the system. 'As we pass [two degrees], it is no longer possible to grow grains at scale in the centre of Russia and North America, where temperatures will increase twice as fast as the global average. Millions will starve, tens of millions of climate refugees will be heading in our direction, and the world economic system will collapse. We are hurtling towards this moment of truth.'

I'd rarely read anything as stark. I knew that Hallam's prediction of grain crop failures the moment we passed two degrees was too precise to be the whole truth as climate scenarios always have error bars attached. But I also knew that he had outlined just one of the many scenarios leading inextricably to irreversible misery for millions in the coming decades and annihilation by the end of the century. Arctic and Antarctic ice melt, rising sea-levels, salination and desertification of farmland and lethal weather events had all entered my peripheral vision but I hadn't managed to derive the equation of 'CLIMATE CHANGE = MASS MURDER' (an Extinction Rebellion slogan), much less apply it to those I loved. Now I had.

Back on Westminster Bridge I was still waiting for 'the signal', whatever that was. People were arriving – some singly, some in small groups – and loitering on the walkways either side of the road. I'd made my way three-quarters of the way across, towards the Houses of Parliament. Next to me three mild-mannered middle-class women, two in their twenties, one in her fifties, chatted about how low-level but wanton lawbreaking wasn't normally their thing. I silently empathised.

We waited. Traffic on the bridge became patchy. Then it stopped. Ten, fifteen minutes passed. Just when I was starting to wonder if the whole thing was off, a woman in a yellow steward's vest strode out into the road, swung her arm up over her head and whistled impressively. We all clambered over. People cheered and hugged. By now one of the few, if not the only, person not in a group, I whooped and clapped by myself, self-consciously.

We gathered and sat down on the cold tarmac for the speeches, delivered from a small platform with the help of a single loudspeaker balanced precariously on top of a stand. After about twenty minutes four police officers picked their way through the huddled mass, doing their best police officer impressions: 'OK, we need to clear the bridge now. You all need to move on.' Nobody moved. Soon after we were told that Extinction Rebellion had blocked all five bridges as planned. Another hour went by and the police concentrated their arrests on London's other, less populated, bridges. That meant we were able to stay on Westminster Bridge for the afternoon.

I was silent the whole time I was on the bridge. I'd dropped into a mood of attentive detachment, in a reverie that somehow did not shut out what was going on. There was a queue to speak on the platform, and anyone could join it. Someone read out their own Kipling-esque poem lamenting the destruction of nature. Another talked about alternatives to gas boilers. Several people did little more than express their exasperation at the endless abuse of the living planet.

The speech I can't erase from memory was from an old man, not so steady on his feet. This was Mayer Hillman, whose name I recognised as one of climate science's elders. 'I'm here to give my support for what you're doing,' he said croakily but confidently. 'But I'm asking you to be realistic.' By which he meant accept defeat. India and China are developing and carbonising, he said, and there was nothing to do about that. Try and minimise your carbon footprint, he advised, but don't imagine you can buy the Earth much extra time. The game is up.

After two or three hours of sitting in the cold I decided to stretch my legs. There were about three or four hundred of us on the bridge and it was largely empty. Once out of earshot of the speeches, all was calm but for the tourists streaming down the pedestrian walkways on either side. I joined them, bobbing along like a cork in a river of humanity towards London's Southbank, where I joined thousands more people, all oblivious to the protests nearby, all buying and buying and buying from the shops, bijou market stalls, chain restaurants, fuelling the great machine.

An obscure record I'd come to know during the time I'd spent in Paris in my twenties started playing in my head. Over a snaking double-bass figure tracked by a hoarse saxophone, with African percussion clattering all around, the French-Togolese actor Alfred Panou laid out where he was at with French society.

'Je n'ai plus le moral nécessaire de penser gauche et d'agir droite.' (I no longer have the moral fibre to think left and act right.)

I headed back to Westminster Bridge, having lost the moral fibre to stay undercover.

2.

The next morning, I felt an overwhelming desire to go to a Quaker meeting. It had been three years since my last visit. I felt this desire at a purely emotional level, as a force acting somewhere in my chest. I rationalised it. It would be good, I reasoned, to reconnect with the movement that had shaped me so that, twenty-two years after I'd left school, I ended up among the few hundred on the bridge and not the many thousands at the Southbank.

I'd spent the night at my in-laws and, as it happened, they lived five minutes from a Quaker meeting house. Yet instead I followed a magnetic pull towards central London and Friends House, the Quakers' imposing administrative centre in Euston, forty-five minutes away.

If people know one more thing about the Quakers, other than the pacifism, it's normally that their religious services take place in silence. In fact meeting for worship isn't just about sitting in silence. When someone, anyone, feels strongly moved to speak, they can, and on any subject whatsoever. This spoken ministry tends to be brief, and it's rare that more than a handful of people will minister during a meeting.

That morning, the first speaker was on their feet after five minutes, also a rarity. She spoke about Extinction Rebellion. The time for signing petitions, for going on marches, for concentrating purely on individual lifestyle choices had passed, she reckoned. The world that sustains us is on the verge of collapse. It is time for more direct action. She hoped that Extinction Rebellion would 'continue to be spirit-led' and could deliver the action needed. She sat down.

Wow, I thought. Whenever I'd brought up Extinction Rebellion with people I knew, they'd never heard of it. The previous night on the TV news, the bridges protest was given thirty seconds. Would anyone else who wasn't on a bridge yesterday be talking about it right now?

A man stood up minutes later, reminiscing about walking through the woods with his mother who told him that the world was God and, in a way, he had never left that view. And then every five minutes or so, people stood and talked about the desecration of the living world and their hopes for Extinction Rebellion. I was astonished.

It was the most vocal meeting I have ever attended. In the gaps between the ministry I wondered if I was willing to be arrested for this movement, face time in jail even. The criminal justice system was utterly unknown to me and I feared it.

With the meeting nearing its end someone stood up and began reading from their phone:

'Corder Catchpool served in the Friends Ambulance Unit during the First World War, but on the introduction of conscription he returned to England to give his witness as a conscientious objector and was imprisoned for more than two years...'

We heard Catchpool's statement from his trial. He longed to return to the ambulance unit.

'By the feverish activity of my hands, I might help to save a fraction of the present human wreckage. That would be for me no sacrifice. It costs far more to spend mind and spirit, if need be, in the silence of a prison cell, in passionate witness for the great truths of peace. That is the call I hear.'

He honoured, he said, those who followed their conscience and had gone to fight.

'In a crisis like the present it would be unbecoming to elaborate the reasons which have led me to a course so different. Today a man must act. I believe, with the strength of my whole being, that standing here I am enlisted in active service as a soldier of Jesus Christ, who bids every man be true to the sense of duty that is laid upon his soul.'

Wow again. Normally when I'm reading or listening to someone speak, I have my guard up against anything that even comes close to referencing a higher power, let alone a specific deity or son thereof. Any mention of such things and I can safely knock the whole argument down. The deities were man's attempt to make sense of the world before science came along and showed, definitively, that there was no sense to the world, that everything was just stuff. I didn't do God.

But my guard was down and these words had hit me hard. They had somehow answered my questions without addressing them directly. They had, to use a Quaker expression, spoken to my condition.

On my way out I was asked if I'd like to stay for tea. I gave some dry-mouthed, garbled reply about having to be somewhere and left. What had just happened?

I still don't quite know. I have learnt, though, that the support for Extinction Rebellion given voice at that meeting was no fluke. It's not just that lots of Quakers cottoned on to Extinction Rebellion in the early days. XR is, I'd argue, the latest in a line of secular campaigning organisations including Amnesty International, Greenpeace and Oxfam in which the formative impact of Quakerism was essential.

Both of the ideological architects of XR – Roger Hallam and Gail Bradbrook – were strongly influenced by Quakerism: Hallam via an engagement with the peace movement in the 1980s and Bradbrook as someone who first attended Quaker meetings as a child and remained a Quaker well into adulthood. Rupert Read, the University of East Anglia philosophy professor and XR spokesperson, is a Quaker. So too is Molly Scott Cato, the Green MEP who was one of XR's earliest political supporters.

And unlike its Quaker-driven forebears, which are mostly conventional organisations acting on matters of Quaker concern, XR has Quaker precepts baked into its operational DNA. They're everywhere from the non-negotiable commitment to non-violent activism to the non-hierarchical structure and collective, consensual decision-making.

Back in November, walking shell-shocked out of the meeting, I didn't know any of this. I just knew that I'd had what could best be described, even by a person as irreligious as me, as a religious experience.

I returned the following Sunday. Soon afterwards I started attending meetings closer to home and now I'm part of that small community. In the months leading up to the April 2019 protests, my local meeting was the one place I could go and talk about the climate crisis and XR and be immediately understood. I could also voice my willingness to be arrested and for that not to elicit looks of incomprehension and mild horror. Non-violent civil disobedience is an accepted way for Quakers to live their faith.

What has provoked more curiosity among Quakers, bizarrely, is my status as a former pupil of a Quaker school. 'Quakerism must be at the core of who you are, then,' a Quaker friend suggested one morning before meeting for worship. At the time, I rubbished the idea. My parents weren't Quakers and didn't know much about Quakerism when they sent me to the school. My school wasn't even that different from any other, most of the time; we sat in class, we did homework, we messed around when we could. I wasn't spending my free time poring over *Quaker Faith and Practice* (the movement's defining text). I was listening to rock and rap music on the radio, vegging out in front of the TV, chatting teenage inanities with my friends. Those were my formative influences.

But, no, this wasn't right. I was one of the many young people who, as the fog of childhood clears, start to apprehend adult society as a vast morass of cruelty, injustice and hypocrisy. I lived in a country permanently at war while proclaiming peace, driven by greed while vaunting charity, avoiding all existential reflection while self-congratulating on its intellectual advancement.

Having held fast to its central commitments to peace, equality, simplicity and integrity for 350 years, Quakerism resonated deeply with me. I knew, though, that if I hadn't gone to a Quaker school then some other dissident worldview would have captured me. Marxism and far-left politics might have appealed. Or, if I'd been born about twenty years later, I could see myself going from YouTube sermon to fundamentalist chatroom, along the path of radicalisation to who knows where?

My Quaker friend made her comment about Quakerism being at my core in February, when Shamima Begum, the young British woman who left the UK at age fifteen to join ISIS, was back in the news asking to be allowed to come home.

The following week in meeting for worship I wanted to stand up and say: 'I am Shamima Begum.'

3.

On 16 April 2019, I stepped once again onto a London bridge occupied by Extinction Rebellion and into a very different experience. It wasn't the same bridge – Waterloo Bridge this time – and I wasn't the same man but it was other factors that made the difference.

In November when I returned to Westminster Bridge after my trip along the Southbank I had the sense of rejoining a sideshow, the kind of minor protest rally you might come across in London on a weekend, look at for a few minutes and then move on. This time I was joining London's main event.

It was the second day of the protests, a Tuesday. I'd spent the morning at the Marble Arch roundabout, normally a choking sink-hole of traffic fumes, but now occupied by XR and entirely traffic free. My XR 'affinity group' (a team of around ten people who undertake civil disobedience actions together, with each person fulfilling a different role) had been assigned to the triumphalist monument called Marble Arch before the week began.

According to the affinity group rota we were supposed to be on shift, manning the roadblocks, but weren't really needed. 'Section 14' orders had been issued for the three other sites – Waterloo Bridge, Parliament Square and Oxford Circus – meaning that protesting at those places was an immediate arrestable offence. Marble Arch was not under Section 14 and was therefore relatively safe. Groups of protesters still blocked the roads branching off from the roundabout but only enough were needed to ensure the police didn't try to push us back and shrink our territory.

I'd come to Waterloo Bridge with Rob, a Quaker who I first met at Westminster Meeting House, not far from Trafalgar Square, which I'd also started visiting. Rob was the only member of my affinity group to have turned up on shift that morning. The others had mostly taken up other jobs within XR: one as a driver, another on first aid, and one person had joined the Samba band. As someone willing to be arrested, I hadn't volunteered for any extra jobs. I saw my job as blocking roads and getting arrested at Marble Arch. The idea of walking into the arms of the police wherever the need was greatest didn't appeal.

I'd started the day in a sour mood. XR's efforts the day before, which included blocking off the terminally congested crossroads at Oxford Circus by planting a bright pink boat complete with DJ booth in the middle of it, had garnered relatively meagre coverage, including another 30-second slot on the night's BBC News bulletin. What would it take to get the greatest threat to human existence ever encountered a little higher up the news agenda?

The high spirits at Marble Arch were infectious, though, and my mood soon lifted. The previous night XR had held on to Waterloo Bridge despite a concerted effort by the police to clear it. The police had run out of holding cells after making 122 arrests and were shipping arrestees as far afield as Brighton to be booked. A conversation with a police officer that morning was relayed around the site. 'You broke us last night,' the police officer had apparently said. Ha! We broke the police!

Really though? As Rob and I stepped onto Waterloo Bridge, my mood suffered a sharp reversal. A row of police vans lined one side of the bridge attended by a sizeable pack of police officers. After a chilly morning Waterloo Bridge was now roasting under an unseasonably intense sun and the police officers' yellow vests appeared oddly, painfully garish, like a detail from a nightmare.

There were cheers and whooping from the crowd further up on the bridge and four police officers came into view carrying a young man by his arms and legs, his face red from the heat. He was dumped down on the pavement by the side of the bridge, looking distraught. The police officers regrouped. They chatted and then another pack strode up the bridge, their faces hardened against us, to where about 40 XR protesters sat in the full glare of the sun.

Hundreds of other people were gathered nearby on the south side of the bridge, many of them around a lorry trailer that was doubling as a stage for impromptu performances and speeches. These people – Rob and I would join them – were all breaking the law, but the police returned time and time again, always moving in packs, to fish someone out at random from the area in front of the lorry.

For the last month I'd been signed up to a range of online updates from environmental campaign groups and news services. I'd had all the gruesome details of environmental collapse drip-fed into my consciousness and now, watching these arrests, they started overflowing: the studies on Arctic ice melt (14,000 tonnes per second), the mass extinctions of wildlife (60 per cent drop in wild vertebrate populations in my lifetime), desertification (12 million hectares of arable land lost annually); the absolute certainty of human suffering and death on a scale never before witnessed.

What did the police think they were doing, exactly? Was a liveable future an unreasonable demand? Was there any other way to achieve it? Didn't the police wonder if they might be in the wrong, arresting that young man barely in his twenties for whom a criminal record would be a longstanding blight, but was still sitting there patiently? How about when they helped that grandmother to her feet and carted her off to the police van? Aren't these the people they are paid to protect?

A sense of injustice overwhelmed me and I wasn't keen to stick around. Rob and I returned to the underground station and said goodbye. He went home to nurse a throat infection and I headed to Oxford Circus.

The party was in full swing. House music thumped out of the pink boat that XR had parked at the intersection of two of London's busiest shopping streets. All around people threw up their arms and danced in the sunshine like revellers at an all-day rave in a remote field in Gloucestershire. Except this was Oxford Circus and tourists were streaming in and out of NikeTown, H&M, TopShop carrying plastic bags shining slick and oily in the sun, staying on the pavement even though the road was clear, hitting the shops like normal, as if a bright pink yacht with 'TELL THE TRUTH' written on the side and a 40-foot mast was something you saw every day in Oxford Circus, maybe part of some megabucks promotional event and we were all happy to party it up with our favourite brand.

A man in his twenties stood up next to the boat DJ, fiddled with his mic and the music came down a notch. I can't remember the exact words, but they went something like this: 'We need to remember why we're here. We're here out of love for the living world. Which is being destroyed. We act in solidarity with those peoples all around the world who are already suffering and dying because of this. We are now in open rebellion [cheers from the crowd] against governments around the world, and our government, for failing to protect us, the people, from this catastrophe...'

In the shade, under the hull, sat the 'barnacles' – people who were joined to each other via 'lock-on' tubes or stuck to the boat, or the trailer it sat on, with glue.

I'd arrived during a lull in arrests and only a few police were present. Had it not been for their hi-vis vests and flak-jackets, they could have been mistaken for bystanders, looking on and wondering what all the fuss was about.

With no-one to talk to, I sent a text to see if any of my affinity group were around and headed down to Marble Arch. Once there, I met up with Laura, the group's Samba band member. Both abuzz with what we'd witnessed over the last two days, we sat and talked a-mile-a-minute while a folk band played on the lorry-trailer stage. Her phone pinged with a message from the site chatgroup. The police had increased their presence at the roadblock on Edgware Road and reinforcements were needed. We headed over.

A crowd of twenty or so XR protesters were blocking the right-hand lane of the road while four or five officers stood around a police car immediately to our left watching us. I recognised a couple of people I'd met that morning holding up a long vinyl banner carrying the slogan 'LIFE OR DEATH' in fat white letters on a black background. I joined them and we started chatting like work colleagues catching up after a long weekend.

More reinforcements arrived. We chanted some slogans and tried to learn some XR songs. At some point someone brought over a Bluetooth speaker and we sang along with Bohemian Rhapsody. As it turned dark, one of the site coordinators doled out hummus sandwiches to the banner-holders. People introduced themselves and struck up conversations without awkwardness, as if in a movie-fantasy London. Many of us had never been part of any movement or political party or activist group before. All of us, I think, had stared into the abyss that was claiming lives all across the equator and would claim ours, too, or our children's. We'd not really spoken of this to many people before; it had been our secret shame. But now, incredibly, we'd found each other. We hugged when we said goodbye.

I divided my time over the next two days between Marble Arch and Oxford Circus, which also became a site for mass arrests. A pack of police would walk up from the vans parked on the north side of Regent Street, move through the crowd, who may be dancing, standing or sitting, and surround someone at random.

'Can you hear me? Do you speak English? OK. There is a Section 14 order served on this site meaning that if you stay here you will be arrested. If you want to protest go to Marble Arch, but you will be arrested if you stay here. Do you understand?'

Fear tightened my throat each time they approached. Would I be next? Would I keep my cool if I was? Would I fold and head down to Marble Arch? Or would I draw on my bubbling reserves of anger and resentment and start getting testy, aggressive even, as all the arrestees before me hadn't done?

I never found out. I didn't stay at Oxford Circus for much more than an hour at a stretch. I wasn't arrested.

My 'Rebellion Week', as it had originally been billed, ended early on Friday afternoon as family commitments called. I returned to central London on Sunday for meeting for worship at Westminster Meeting House.

I smiled as I sat down in the large wood-panelled meeting room, as if it were a concert and I was waiting for a favourite singer to come on stage, and belt out the hits that had wowed me back in November. They never came, but over half-way through one of the meeting's regulars rose to voice their support for XR. This was hesitant, careful ministry, and all the more affecting for that.

It was followed by ministry from a Quaker visiting from Italy, who was similarly enthused by XR's actions and, after that, from an XR activist attending his first Quaker meeting. Five minutes later a rangy man with close-cropped hair and a northern accent stood up and observed how the injunction 'attend to what love requires of you', a phrase much beloved by Quakers, had been at the heart of everything he'd witnessed during the XR protests. He had seen XR protesters acting out of love for the police and their families, even as they were arrested. Thanks to this, he said, he had come to a deeper appreciation of non-violence. It meant more than 'just not hitting the police'. Acting non-violently, with love, allowed for much greater possibilities than that.

We got chatting over tea after the meeting. During the previous week, coverage of the XR protests had grown steadily so that climate change and ecological collapse were finally being discussed in honest terms in the mass media. This was an incredible feat in such a short space of time, I said. XR had exceeded his expectations too and he almost felt overtaken by events, he said. How long had he been involved, I asked. 'I'm one of the co-founders,' he replied cheerfully.

His name was Ian Bray, a Quaker from Huddersfield. He had met Gail Bradbrook and Roger Hallam in one of the activist groups that preceded XR when Bradbrook and Hallam were in the early planning stages for the movement. 'I didn't have to try and import Quaker values into XR,' he told me. 'They were already there.'

I'd found myself, quite unexpectedly, at the centre of something very big. For the last twenty years I would mention the Quakers and elicit only vague recognition or jokey references to porridge. For the last six months I would drop Extinction Rebellion into conversations and draw a blank.

Now these two groups' values and ideals, which are so closely linked, were setting the agenda.

I left the meeting house carrying the intoxicating feeling I was riding a great wave which would build and build until it carried the whole world along with it.

4.

The following week, it broke. By Monday all of XR's sites apart from Marble Arch had been cleared of protesters. On Tuesday, I returned to my computer at work and heard again the steady drip, drip, drip of dismal environmental news as it landed in my inbox or Twitter feed.

On Thursday, the day after the last protester had been cleared from the roadblocks at Marble Arch, I left work to join the closing ceremony, which was quickly rebranded a 'pausing ceremony', for the Rebellion Week, which had lasted eleven days.

It was a beautifully clear and slightly chilly evening. Several hundred XR supporters sat on the grass by Speakers' Corner in Hyde Park, just across from the Marble Arch roundabout. I arrived late and joined the standing circle of people at the edge of the seated crowd.

The powerful solar-powered PA rig that XR had used at Marble Arch had vanished and, in an echo of the Westminster Bridge protest, had been replaced by a single speaker, which three people held above their heads. A woman spoke in a calm voice about feeling grateful for everyone who had joined in the XR protests, for the police, for the patience of the Londoners who'd had their lives disrupted. She spoke of her love for the world and how we had to rebel to protect it and just as it all became a bit New Agey, the woman next to me pointed up at the sky and a plane on its descent into Heathrow with another behind it visible in the distance.

'Every three minutes,' she said sadly. 'Every three minutes.'

A young Muslim woman led the crowd in a trance-like repetition of two elements of the Adhan, the call to prayer, before Helen Burnett, a Church of England Deacon who I'd met at Marble Arch on Tuesday morning, delivered a sermon, and a Rabbi gave a short speech.

Rebellion Week had begun with a multi-faith service – led by Rowan Williams, the former Archbishop of Canterbury, outside St Paul's Cathedral

– and now it was ending with one. This seemed right, considering the crucial role that religious groups had played throughout. Churches and meeting houses and temples (and doubtless other places of worship I was unaware of) had hosted XR protesters from outside London and many kept their doors open to activists in need of a rest throughout the day.

After the religious speeches, things turned vague and New-Agey again and I decided to head off, feeling intensely sad. Above me the planes continued their descents, pumping untold tonnes of carbon dioxide into the air. Every three minutes, like my neighbour said, every three minutes.

Had XR made any difference? Would people stop buying all the mountains of crap they didn't need now? Ditch their cars and business trips and smartphones? Would they stop wanting these things? Would the rainforests no longer be burnt to make way for oil palms? Gas pipelines be turned off? Plastics factories shuttered?

I reached the pedestrian crossing leading to the Marble Arch traffic island and my sadness gave way to anger. A week ago, I'd stood here when it was traffic-free and the air was clear. Now it was thick with fumes. There are up to 60,000 premature deaths a year in the UK due to air pollution, one recent study estimated. Sixty thousand people, each year!

I crossed the road and felt the rage welling up inside me and wondered if I could contain it. This sometimes happens as I walk up the long main road leading to my flat on my way home from work. I think about these public health and environmental studies and the human misery they express in cold data, and yearn to get even, to commit some outrage.

The main road is always chock-a-block. Always. And there's normally just one person in each car. Maybe they make the same journey every day, and don't care that they are contributing to 60,000 early deaths a year, or 14,000 tonnes of Arctic water per second flowing into the sea, or carbon dioxide concentrations reaching levels not seen since the Pliocene era when forests grew at the North Pole and sea levels were 20 metres higher than today.

How do these lone daily drivers think all this will pan out? How will we escape the catastrophes: the heatwaves, the floods, the starvations? Do we expect the laws of nature will bend to accommodate us? That we can throw our children off a cliff and they will not hit the ground?

What? Were you expecting a happy ending?

DENIAL, DECEIT, BULLSHIT

Gordon Blaine Steffey

This is the most deceptive, vicious world. It is vicious, it's full of lies, deceit and deception.

Donald J. Trump, POTUS, *60 Minutes* interview (2018)

Bullshit is unavoidable whenever circumstances require someone to talk without knowing what he is talking about.

Harry Frankfurt, Philosopher, *On Bullshit* (2005)

The English word 'denial' had a life before the Freuds. Life since has been emphatically coloured by psychoanalytic theory, where denial names what first generation Polish Freudian Hermann Nunberg efficiently termed a 'psychological annihilation of reality.' By all accounts British environmentalists George Marshall and Mark Lynas penned the first reference to 'climate deniers' in a 2003 jeremiad for the *The New Statesman*. They acknowledge the influence of British sociologist Stanley Cohen's *States of Denial* (2001), a far-reaching work invoking papa Freud and the denial mechanism to examine states of mind and cultures 'in which we know and don't know at the same time.' While Cohen uses denial to fathom individual and collective disavowals of atrocity and suffering, Marshall and Lynas use denial to spotlight 'patterns of behaviour' reeling violently out of step with ostensible acceptance of the scientific consensus on climate change (minimisation or 'soft' denial). Such denial haunts us all. The loan from the Freuds via Cohen stuck, perhaps because the dialectic of denial (what Cohen loosely terms knowing and not-knowing) best interprets across stakeholders the staggering persistence of 'business as usual' and/or the astonishing deficit of urgency in the face of atmospheric CO_2 on a steep climb to 500 parts per million inside 2050. This would mean a global temperature increase of at least 2° Celsius. No members of genus *Homo* have seen such a world and its current members will wish they had not.

Climate is critical terrain in the enduring polarisation of America, the roots of which lie in the mounting strife between fundamentalists and modernists at the turn of the twentieth century. The two camps engaged fatefully on the matter of evolution instruction in high school biology classrooms and textbooks. Violation of Tennessee's antievolution statute culminated in *The State of Tennessee v. John Thomas Scopes* (1925), that storied high noon between celebrity litigator Clarence Darrow and former US Secretary of State William Jennings Bryan. To the American Civil Liberties Union, *Scopes* was a defence of freedom of speech and the establishment clause of the First Amendment, which would not technically apply to states until *Cantwell v. Connecticut* (1940). To Darrow, *Scopes* was an attempt to assail superstition in the name of science. At Dayton he thundered away at Bryan: 'You insult every man of science and learning in the world because he does not believe in your fool religion.' All know the outcome: a technical and temporary win for the fundamentalists (a guilty verdict and £79 fine for *Scopes*) and a colossal win for science in the public square. Creationist activity since the 'monkey trial' tilts away from the biblicism of fundamentalist Protestants and pools in the critique of scientific materialism from the vantage of 'intelligent design' or ID. Historian Edward Larson characterises ID partisans as reduced to 'discrediting the theory of evolution by doggedly looking for gaps,' a blueprint sketched by Bryan in his closing remarks at Dayton: 'Evolution is not truth; it is merely a hypothesis – it is millions of guesses strung together.' When the evangelical right roared back into the public square in the 1970s and 1980s, evolutionary biologist Stephen Jay Gould complained that 'nothing has changed' since *Scopes*. Primitive and latter-day creationists mischaracterise routine scientific uncertainty as symptomatic of mere guessing and inconclusive debate in a 'rhetorical attempt to falsify evolution.'

US courts have so far protected public schools against creationism in whatever state of dress by invoking the establishment clause of the First Amendment (the same clause sourcing the temporary restraining order issued by US courts against the 2017 'Muslim ban'). Anti-evolution statutes of the last century have yielded to irregular efforts to undermine evolution by, say, authorising non-specialist critique of settled science in classrooms (as in a 2019 bill introduced to the South Dakota house) or finessing the margins between theory and fact. In 2004 a school board in Georgia resolved to affix a disclaimer to biology textbooks: 'Evolution is a theory, not a fact, regarding the origin of living things. This material should be approached with an open mind, studied carefully, and critically considered.' A co-author of the stickered textbook testified at the ensuing trial that the disclaimer exploits an American vernacular where (as Stephen Jay Gould puts it) '"theory" often means "imperfect fact."' Gould nevertheless insisted on the *fact* of evolution where fact means '"confirmed to such a degree that it would be perverse to withhold provisional assent."' Biologists from Ernst Mayr to Richard Dawkins have insisted that explanations for the fact of evolution should not obscure the fact of evolution, which, in the words of Theodosius Dobzhansky, will be 'doubted only by those who are ignorant of the evidence or are resistant to evidence, owing to emotional blocks or to plain bigotry.' Gould grieved the politics driving the late twentieth century resurgence of a religious right who resorted to cynical tactics of 'distortion and innuendo' in their campaign against the fact of evolution. He was also 'saddened' by an emerging propensity to mute legitimate scientific debate about evolution in overreaction to crusades to malign sound science for ideological ends.

The parallels to climate denial—where stakes are incomparably higher and the belligerents profit from better organisation, resourcing, and strategic communications—are imperfect and instructive. The playbooks bear a family resemblance: scientists camping indignantly on the science and scientific consensus in the face of pseudo-scientific dissent motivated by values and commitments extrinsic to the science seemingly but not actually in question. In *Evolutionary Biology* (1986), Douglas Futuyma observes that 'no biologist today would think of submitting a paper entitled "New evidence for evolution"'—why? 'It simply has not been an issue for a century.' In her seminal paper 'Beyond the Ivory Tower: The Scientific Consensus on Climate Change' (2005), historian of science Naomi Oreskes used keyword sampling to analyse article abstracts published over a decade in 8,500 refereed science journals on the topic 'global climate change.' She finds that 'the basic reality of anthropogenic global climate change is no longer a subject of scientific debate,' a finding consistent with white papers from NGOs and professional organisations like the National Academy of Science, American Meteorological Society, American Geophysical Union, American Association for the Advancement of Science, US Global Change

Research Program, Intergovernmental Panel on Climate Change, and so on until dawn. Oreskes' findings have been reproduced in two studies published by the refereed *Environmental Research Letters*. A 2013 study found that 97.1% of actively publishing scientists accept anthropogenic climate change and a 2016 study synthesised six independent analyses of the state of consensus, finding that 90–100% of actively publishing climate scientists accept anthropogenic climate change. Despite this state of consensus, despite foot-stamping verdicts of 'the science is clear' and 'the debate is over,' many Americans aren't getting the brief.

Faint cheer may be built atop a 2018 report from the Yale Program on Climate Change Communication: a strong majority of Americans (73%) accept that 'global warming is happening' and a mounting majority (62%) that human activity caused and causes the chemistry of our atmosphere to change. Curiously, the same report finds that only 57% of Americans believe that most scientists think 'global warming is happening' (cause notwithstanding). One in four Americans imagine 'a lot of disagreement' about global warming in the scientific community and a meagre one in five know the scientific consensus on anthropogenic global warming sits north of 90%. Whence the gulf stretching between Americans' notions of what scientists think and what scientists actually think? Does uncertainty about what scientists know bear on support for adequate (much less progressive) climate policy? While science outreach and collaboration with entertainment media have given scientists a wanted makeover – recall the plucky botanist from *The Martian* (2015) whose survival depends on having 'to science the shit out of this' – I wonder whether we have thought enough of the consequences of scientists constituting a 'new priesthood.'

In evolutionary biology robust debate about the 'mechanism of evolution,' framed by George Gaylord Simpson as debate about the rate and manner of evolutionary change or 'tempo and mode,' was misrepresented by creationists as controversy about the fact of evolution. In an opinion paper on climate change, biologist Stuart Firestein argues that the corrigibility of science renders it vulnerable to 'distortion and innuendo' by mercenaries and ideologues seeking to sandbag the validity of scientific knowledge for capital. Physicist Niels Bohr's shibboleth to the effect that every utterance is a question does not imply brute scepticism, but rather affirms philosopher Karl Popper's elegant conclusion to *The*

Logic of Scientific Discovery (1959): 'Science never pursues the illusory aim of making its answers final, or even probable. Its advance is, rather, towards an infinite yet attainable aim: that of ever discovering new, deeper, and more general problems, and of subjecting our ever tentative answers to ever renewed and ever more rigorous tests.' When the scientific community woke to the gravity of climate change denial and the need to respond to it, deliberations tended to deteriorate into executive repetition of the science or the 'blame game.' On whom or what shall we settle blame for denial? On the fossil fuels industry? On the ill-fit of science and the exigencies of media? On the scientific illiteracy of the American public or perhaps on the maladaptation of our brains to ambiguity? Candidates for the guillotine were abundant.

An influential study tracks climate disinformation to a cadre of anti-regulatory scientific outliers who colluded with nationalist and corporate interests to gin up confusion about sound science in service of free market capitalism, seen as the sole viable contestant in the cold war against global communism. Such is the thesis of Naomi Oreskes and Eric Conway's *Merchants of Doubt* (2010), which shares DNA with other environmental and climate exposés from *Silent Spring* (1962) to David Michaels's *The Triumph of Doubt* (2020). The title *Merchants of Doubt* refers to an unattributed 1969 proposal circulated internally at Brown & Williamson Tobacco, which reads in part: 'Doubt is our product since it is the best means of competing with the "body of fact" that exists in the mind of the general public. It is also the means of establishing a controversy.' The problem in view is how to contest the public consensus that 'cigarettes are in some way harmful.' The solution is to create a semblance of unsettled debate through dispensation of doubt, a template that has been adopted lock, stock, and barrel by the climate disinformation complex. The latter names a nexus of government, media, corporate, and independent actors who generate doubt about sound and settled science in order (at a minimum) to slow down regulatory and policy processes pertaining to the extraction of fossil fuels and carbon emissions. The bottom-line is *the bottom-line*. 'Distortion and innuendo' are the daily bread of this nexus, which dwarfs in power, organisation, and efficiency the fundamentalist Protestant base backing Jehovah against Darwin. The fundamentalists were 'true believers' striving against a godless worldview that flung wide the gates to decadence, but internal correspondence like the proposal above and memo below express something other than true belief, something more akin to depraved indifference and mercenary cynicism. In a now infamous memo to the George W. Bush White House, Republican strategist Frank Luntz advised: 'Should the public come to believe that the scientific issues are settled, their views about global warming will change accordingly. Therefore, *you need to continue to make the lack of scientific certainty a primary issue in the debate*' (emphasis in original). Luntz supposes that my view of the scientific consensus on global warming bears strongly on my view of global warming. A 2015 study by van der Linden *et al.* confirms that public perception of the scientific consensus is a 'gateway belief' and that improving this perception is 'significantly and causally associated with an increase in the belief that climate change is happening, human-caused and a worrisome threat,' which beliefs are in turn predictive of support for actionable public policy to abate the threat. Again, only 20% of Americans have any inkling about the actual state of the scientific consensus on anthropogenic global warming. The gateway has been obscured by the distortion and innuendo of death-dealing merchants of doubts.

The climate disinformation complex undermines science relevant to public policy by circulating uncertainty for our consumption and confusion (#Exxonknew). That hashtag tracks to a growing movement hellbent on holding ExxonMobil accountable for climate deceit. A wide paper trail suggests that ExxonMobil knew of the adverse effects of CO_2 emissions on our atmosphere and climate as early as 1981, seven years before National Aeronautics and Space Administration scientist James Hansen testified to a Senate committee that 'it's time to stop waffling' and insisted on the causal link between anthropogenic release of CO_2 and climate change. In a mantic mood, Hansen published a paper that same year in *Science* where he projected warming of around 2.5°C by the end of the twentieth century. Meanwhile, bootlegging the Tobacco playbook, ExxonMobil denied publicly what they accepted privately and bankrolled blue-chip climate deceivers like aerospace engineer Willie Soon as late as 2010. Science historians Supran and Oreskes studied climate change communication at ExxonMobil between 1977 and 2014, finding that while scientists on its payroll contributed constructively to the advance of climate science, ExxonMobil 'promoted doubt about it in advertorials [and] misled the public.' Specifically, while 80% of internal communication traffic accepts anthropogenic climate change, 81% of advertorials (advertisements formatted as editorials) express doubt. As the dissemination of doubt by Tobacco exacted a terrible tax in human lives, so too will the doubt proliferated by climate deceivers and misleaders levy an incalculable cost in human and non-human lives, cultures, and environments.

The cost is already being levied on the climate frontlines among small island, rural, and indigenous peoples largely invisible to the American in the street. The vulnerability of the Marshallese in the central Pacific, the pastoralist Gabbra in northern Kenya, and Inuvialuit communities in the western Canadian Arctic is a corollary of their geographies and/or their resource-based lifeways. An eye to communities where the graver impacts of climate change are already afoot may serve to wake up those whose own vantage enjoys more security for now. More economically developed countries will shoulder responsibility for sheltering climate frontline communities within and 'climate refugees' from beyond their territorial borders, especially those from developing and least developed countries which are least able to adapt to adverse climate change impacts for which they have negligible liability. The US is unprepared to shoulder such responsibility with anything in the vicinity of foresight. While the Immigration and Nationality Act of 1965 acknowledged that victims of 'catastrophic natural calamity ... who are unable to return to their usual place of abode' are as entitled to refuge as victims fleeing persecution, the Refugee Act of 1980 eliminated that salutary statutory formulation. In consequence Citizenship and Immigration Services and Homeland Security are provisioned with inadequate and temporary legal accommodations for a class of displaced persons ineligible for refugee status and asylum under current immigration law (unless they are demonstrably fleeing a persecution stipulated by the 1967 Protocol Relating to the Status of Refugees). Stuck in his rut of feverish hyperbole, the American president complains (falsely) that the US has 'weak,' 'pathetic,' and even the 'weakest immigrations laws' of (presumably) developed countries and he sinks administration energies in shrinking legal immigration and asylum, stemming illegal immigration, accelerating deportation and removal, and underscoring our southern territorial border with a wall. In January the Government Accountability Office argued that the administration's rescission of prior Obama-era executive and regulatory actions (which the

White House described as unduly 'burdensome' to the fossil fuels industry) leaves US diplomats unable to respond effectively to destabilising population shifts triggered by climate change, which the national security apparatus characterises as a 'threat multiplier.' US Diplomats must develop country strategies without guidance on 'whether and how to include climate change risks.' The report warned, 'without clear guidance, [the Department of] State may miss opportunities to identify and address issues related to climate change as a potential driver of migration.' In *Global Warming of 1.5°C* (2018), the Intergovernmental Panel on Climate Change (IPCC) insists that 'the worst impacts [of anthropogenic climate change] tend to fall on those least responsible for the problem, within states, between states, and between generations,' an inequity compounded by underrepresentation of the worst-affected-least-responsible in strategic planning for climate change abatement and mitigation. To this puzzle the *Fourth National Climate Assessment* (2018) adds that 'marginalised populations' at highest risk 'may also be affected disproportionately by actions to address the underlying causes and impacts of climate change, if they are not implemented under policies that consider existing inequalities.' In short, the United States is appallingly underprepared to meet climate-related humanitarian challenges not only because momentary distance from the frontlines and climate deceit/denial have stretched and dimmed the pathways leading to humane, progressive climate policy but also because the climate disinformation complex hit pay dirt when Americans elected @realDonaldTrump to head the American state.

Out of the gate, our Twitterer-in-Chief nominated erstwhile ExxonMobil CEO Rex Tillerson to head the State Department. In the thick of his ten-year stint at ExxonMobil, Tillerson characterised climate change as an 'engineering problem' to be solved by unidentified 'engineering solutions,' which will taper the toll of global poverty (and swell the coffers at ExxonMobil) by 'making fossil fuels more available' to the poor. In that same year reality TV prodigy @realDonaldTrump faulted the Chinese for cooking up global warming to sabotage American manufacturing. A turgid stream of innuendo, ignorance, and peacocking followed hard on the heels of this absurdity. Exhibit A is his portfolio of climate tweets belabouring the epidemic confusion of climate with weather—in November 2018 he twittered 'Brutal and Extended Cold

Blast could shatter ALL RECORDS – Whatever happened to Global Warming?' In January 2019 he drivelled 'near record setting cold ... Wouldn't be bad to have a little of that good old fashioned Global Warming right now!' Profuse correction of this blunder by critics and boosters alike hasn't yet discouraged him from recycling it ad nauseam. Meantime he distorts terminology that helps folks differentiate day-to-day atmospheric conditions (weather) from long-term atmospheric trends (climate): 'They only changed the term to CLIMATE CHANGE when the words GLOBAL WARMING didn't work anymore. Come on people, get smart!' Not a trace of irony there, for the record. When he trades innuendo and obfuscation for 'grab them by the p****' candour, we hear the *real* @realDonaldTrump: 'total con job,' 'Global warming con,' and 'This very expensive GLOBAL WARMING bullshit has got to stop.'

A quotation from philosopher Harry Frankfurt's *On Bullshit* (2005) crowns this essay and deserves a digression. Since November 2016 we (Americans) find ourselves sunk in the subject matter more deeply than 66 million of us fancied. A quirk of our constitutional republic termed the electoral college meant that the loser of the popular vote now deluges our White House with the material in question. This is no mere calumny but critical appraisal of the better part of a tonnage of unmistakably dubious statements uttered, tweeted, and otherwise outgassed by this President. In the run-up to the 2016 elections, Frankfurt offered a no-nonsense assessment of candidate @realDonaldTrump: 'It is disturbing to find an important political figure who indulges freely both in lies and in bullshit' (adding that the enthusiastic embrace of such a figure is more disturbing still). Admittedly, it isn't always easy to sort the one from the other, especially when they appear so often alloyed. I offer the following as a test case. The *New York Times* published a 2017 op-ed entitled 'Trump's Lies,' which took the form of a kind of catalogue of deceit. The catalogue includes an observation made in the wake of hurricane Maria in late 2017: 'Nobody has ever heard of a 5 hitting land.' The Saffir-Simpson scale categorises hurricanes by wind intensity and a category 5 (the reference here) exhibits sustained winds of 252 km/h. In fact, by 2017 three category 5 hurricanes had made landfall in the US, two of them well within lifetime of this president. Factcheckers were quick to correct the record, others to expose yet another of 'Trump's Lies,' but is lying the best description of this utterance? It's false, to be sure, and so too his exposition of Maria's impact on Puerto Rico, a US territory: 'It hit right through – and kept to a 5 – it hit right through the middle of the island.' In fact, Maria made landfall as a category 4 and quit the ravaged island seven hours later as a weak category 3. Another lie?

On Bullshit to the rescue. Simply put, a liar deliberately misrepresents what s/he takes to be the truth. The ground is littered with evidence to condemn @realDonaldTrump for lying about many matters over a tremendous range of issues, but perhaps not about this matter. Frankfurt teaches us that the bullshitter presents as a truth-teller while trying to put one over, but unlike the liar s/he is indifferent to the truth-value of what s/he says. The liar cares very much for the truth and seeks to bury it; the bullshitter makes claims about which s/he is unsure and is unconcerned by the fact that s/he doesn't regard them as true (largely because s/he has aims alien to the truth or falsity of the matter in hand). Frankfurt observes that bullshit is inevitable when 'a person's obligations or opportunities to speak about some topic exceed his knowledge of the facts that are relevant to that topic.' After 852 days of this shambolic presidency, we know (surely, all know) that @realDonaldTrump is neither well-informed about those matters of which he is regularly obliged to speak and in which presidents ought to be knowledgeable nor is he driven by curiosity, duty, or shame to make up that deficit. Take the word of his former Secretary of State and ExxonMobil CEO, Rex Tillerson, who confided to mainstream media that the 45th President 'doesn't like to read, doesn't read briefing reports … doesn't like to get into the details of a lot of things but rather just kind of says, "Look, this is what I believe."' As category 5 hurricane Irma bore down on the Caribbean @realDonaldTrump acknowledged 'I never even knew a Category 5 existed' – this was scarcely a month before 'Nobody has ever heard of a 5 hitting land.' Verdict: *bullshit*.

Asked whether the severity of hurricanes Irma and Harvey caused him to reassess his views on climate change, The Donald replied: 'Well, we've had bigger storms than this.' Strictly speaking, this is true, whether you assess the claim by Saffir-Simpson, loss of life, or monetised damages. It is nevertheless bullshit. Frankfurt reminds us that bullshit is what it is by virtue of the indifference of the speaker to the truth-value of what s/he says, not by virtue of the truth or falsity of what's said. Scales tilt toward falsity when the circumstances render bullshit inevitable (as they do here) – @realDonaldTrump hasn't an inkling whether hurricane intensity bears evidentially on climate change. In 2018, he jeered reports of science linking the intensity of recent hurricanes to climate change: 'You'd have to show me the scientists because they have a very big political agenda' (a theme we've encountered before). In any case, he neither knows that nor cares whether 'we've had bigger storms than this' is true – after all, the very same day he said of Irma 'I never even knew a Category 5 existed.' Meanwhile, his 'bigger storms' answer implies that the severity of prior storms spells trouble for the scientific consensus on climate change. In this he exhibits (as he often does) more concern for curating his image and messaging his tribe than accuracy. To be fair, it can be difficult to determine whether he cares about the truth of what he says (though a strong preponderance of evidence suggests he does not). How will you assess 'Mexico will pay for the wall'? On balance, the climate talk of @realDonaldTrump is a potluck of deceit and bullshit uncontaminated by denial in any sense save its flattest. The prognosis for climate futures is dire whether he's lying or bullshitting because policy to abate the cascade of worsening outcomes is already past due.

Trumpers insist the president's views on climate change have thawed. They cite an interview given in the wake of a grim 2018 IPCC special report on the impacts of global warming of 1.5°C above pre-industrial levels. 'I'm not denying climate change,' he opened strongly before lapsing into his inimitable cant: 'I think something's happening. Something's changing and it'll change back again. I don't think it's a hoax. I think there's probably a difference. But I don't know that it's manmade. I will say this: I don't want to give trillions and trillions of dollars. I don't want to lose millions and millions of jobs.' Something ... probably ... I don't know ... I don't want. The thaw (such as it is) distinguishes the something that's happening from a hoax, rejects the scientific consensus on anthropogenic climate change, and jollies the MAGA mob with ostensible regard for taxpayer money and American labour. In fact, he seems not to have strayed an inch from his 2014 tweet: 'Whether Global Warming or Climate Change. The fact is We didn't cause it. We cannot change it.' Perhaps he hasn't wandered terribly far from his 2012 tweet linking global warming and conspiracy. As late as 2016 he explained to bootlickers on the programme FOX & Friends: 'climate change is just a very, very expensive form of tax … I know much about climate change. I'd be – received environmental awards. And I often joke that this is done for the benefit of China. Obviously, I joke. But this is done for the benefit of China.' A word comes to mind …

In 2014 @realDonaldTrump twittered: 'Is our country still spending money on the GLOBAL WARMING HOAX?' America Made Great Again spends increasingly, risibly less than ought to be spent to prevent preventable loss. Notwithstanding prior and planned cuts to global initiatives that run afoul of 'America First' isolationism, federal agencies responsible for energy and environment policy face deep cuts in the president's proposed 2020 budget. The Environmental Protection Agency (EPA) faces a stiff 31% cut in funding, the Department of the Interior a 14% cut, the Department of Energy an 11% cut (felt most keenly by the Office of Energy Efficiency and Renewable Energy and the R & D initiative Advanced Research Projects Agency-Energy), and the National Oceanic and Atmospheric Administration a stout 18% cut (or £790 million less than in fiscal year 2019). Now piloted by a former coal industry lobbyist, the EPA strangely supports the president's proposed budget. In its 2020 budget brief, the EPA explained that 'removing unnecessary regulatory burdens allows the EPA to be a catalyst for economic growth.' This is an abiding motif at EPA under @realDonaldTrump, whose appointees have been given marching orders to shred the 'anti-energy agenda' at EPA. His first appointee has been credited by *The Washington Post* with 'halting the agency's efforts to combat climate change and to shift the nation away from its reliance on fossil fuels.' The 2015 Clean Power Plan has been classed among these 'regulatory burdens' on economic growth and targeted for repeal – it aimed to reduce CO_2 emissions from the power sector (second only to the transportation sector in emissions) by 32% relative to 2005 levels within 15 years. The administration has tied the plan up in political process and litigation since 2016. In an April interview with the *Financial Times*, the current EPA chief confirmed the writing on the wall, ceiling, and floor: 'for all intents and purposes, we have left the Paris climate accord.' If the Furies are just (and the jury is out), this may not be the last word on Paris. In May the US House of Representatives passed the Climate Action Now Act. The bill prohibits expenditure of federal money in service of the White House plan to withdraw from Paris and requires the White House to report to Congress how it plans to meet US obligations under the Paris agreement within 120 days. Republican Majority Leader in the Senate replied that 'this futile gesture to handcuff the US economy through the ill-fated Paris deal will go nowhere here in the Senate.'

Climate Action Now is the first climate bill to be heard in the House in a decade and will not be alone of its kind in 2019. Congressional Democrats have this year convened select committees on 'climate crisis,' cementing the significance of climate change action to the Democrats' legislative agenda in the new Congress and to the 2020 contest for the White House. As the UK Parliament formally recognises the 'climate emergency', the US head of state dispenses darkness and spits bullshit despite the dire foresight on view in the *Fourth National Climate Assessment* (November 2018). The *Assessment* finds that without substantial and sustained global effort to reduce carbon emissions and implement regional mitigation strategies, the United States will suffer substantial damage to its infrastructure, environments, public health, and economic growth. Parenthetically, the president disbanded the committee tasked to evaluate and derive policy guidance from these quadrennial assessments, on grounds of poor industry representation. Findings like these and dissent from the (mis)lead of the White House by critical units and departments of government (NASA, the Department of Defence, et al.) incited @realDonaldTrump to propose for 2019 a President's Committee on Climate Security staffed by denialists, deceivers, and bullshitters. Climate scientist Michael Mann and activist Bob Ward compare the proposed committee to the Stalinist program to 'modernise' agriculture by using the mad science of gene-denialist Trofim Lysenko despite the dissent of reputable geneticists.

If we mourn the miscarriage of the Green New Deal (which was more aspirational than quantitatively rigorous and policy-specific) and the stillbirth of Climate Action Now (which isn't sufficiently progressive, in any case), we take consolation in the inevitable congressional overhaul of the proposed 2020 budget. That budget nevertheless makes plain the state of US executive commitment to climate change action—until Americans return to the polls. Democrats have rightly identified climate as an Achilles' Heel for the White House and congressional Republicans in the

2020 elections and prioritised meaningful climate policy in place of merely green chatter. Just now catching up to public sentiment, Republicans are in the initial phase of formulating a strategy on climate that sits comfortably with the 'old time religion' of subsidies, deregulation, and economic growth. Mann and Ward urge Americans not to be 'conned by the Trump administration's climate Lysenkoism' and rather to 'trust in the robust findings of climate researchers who place the public interest ahead of political ideology.' Put thus, the matter *seems* so simple: either evidence-based public health or party loyalty.

In 2018, Charles C. Mann published *The Wizard and the Prophet*, a tale of two dead men whose views constitute types or 'basic intellectual blueprints ... for understanding our environmental dilemmas.' 'Wizard' Norman Borlaug (d. 2009) believed that proper appliance of science and technology to environmental problems eliminates those problems (a view termed 'techno-optimism'). Borlaug was what Lysenko dreamed of becoming, the architect of a 'Green Revolution' that joined high-yield crop varieties and agronomic techniques to increase global grain harvests and significantly mitigate global hunger (an achievement not without costs, for example, the displacement of smallhold farmers by large-scale or industrial agriculture, soil erosion and compaction, watershed degradation, and fertiliser pollution). For wizards in the type of Borlaug, innovation is the way to banish the besetting devils of climate change. 'Prophet' William Vogt (d. 1968) believed that natural systems seek balance and that human beings must respect ecological limits by radically reducing consumption *or* face globally-scaled devastation (termed 'apocalyptic environmentalism'). His publications expressed contempt for the latter-day idol of economic growth and paired indifference to limits with oracles of doom. This difference between Borlaug and Vogt types sits atop a deep fracture identified by the late Gabriel Marcel. Marcel argued that our modern 'broken world' reduces everything to function. Persons have been reduced to a set of functions (professor, patient, citizen) and timetables for the discharge of those sundry functions (once yearly holiday, physical, tax payments). In such a world the default is to deny 'mystery' and recognise only technical 'problems' to be dissolved by the selection and application of proper technique. Marcel explains: 'a problem is something which I meet, which I find completely before me, but which I can therefore lay siege to and reduce. But a mystery is something in which I am myself involved.' For wizards climate change is a set of corrigible technical problems awaiting the fitting tools and for prophets also a mystery implicating me in incorrigible existential questions of identity, desire, and destiny.

Mann raises a window on climate change deliberation in US public life. In his December 2018 opinion published by the *New York Times*, Republican Senator John Barrasso wears the face of a 'wizard' and incants the political solution of his party to climate change concern: 'The nation is leading the way [?] not because of punishing regulations, restrictive laws or carbon taxes but because of innovation and advanced technology, especially in the energy sector.' Barrasso hails enhanced oil recovery as a case in point, a process where captured CO_2 is injected into oil field reservoirs so as to flush residual oil from subsurface rock formations for extraction, 'producing more American energy and sequestering carbon dioxide underground.' I forbear to explore the mire of the Senator's claim that 'this is good for the environment and the economy,' which originates in the heredity of wizardry. The White House proposal to defund the Advance Research Projects Agency-Energy belies the Republican summons to innovate and lends blood to suspicions that innovation is a politically inexpensive way to concede the necessity of action on climate change without conceding the urgency of aggressive policy action. Vox columnist David Roberts argues that despite promising 'conservative-friendly' climate policy like the Republican-sponsored Energy Innovation and Carbon Dividend Act, Republicans remain protective of the fossil fuels industry, which is essentially 'incommensurate with serious action on climate change.' In Roberts' view, such incommensurability leaves only one route open: 'bullshitting.' By this he means a positive buzz to drown out the roaring abyss of moral complicity with degradation and death wrought by climate change. In a January tweet, Senate Democrat Sheldon Whitehouse offered this genealogy: 'This is such bull***t. First it was a hoax. Then the science was uncertain. Then we were "alarmists." Then any fix was too expensive. Now it will be fixed with "innovation."' Roberts rightly observes that 'market-shaping' policies (and manifestly not *laissez-faire*) generate innovation and yet the historical drivers of innovation (regulating, taxing, and federal R & D funding) are inimical to the current configuration of the Republican party. In short, he unmasks these professed wizards as bullshitters, and pitches his tent with proper wizards when he concludes that 'the climate challenge is simply the challenge of scaling up and substituting clean technologies for dirtier ones.'

Environmental historian Andrew Watson argues that 'techno-optimism is a desperate hope that the problem can be solved without fundamental changes to high-energy standards of living.' He points out that climate models seeking to limit global mean temperature increase to no more than 1.5°C assume a 'continuity of energy consumption' across the twentieth century and are therefore complicit with energy poverty and 'incredible energy inequality' between industrialised/industrialising and non-industrialised countries. Watson cites data from the International Energy Agency epitomising the trouble: industrialised countries in North America and Western Europe consumed three to four times as much energy per capita as the global average in 2016. Watson argues that 'technology will not save us,' and that only 'dramatic reductions' in energy use will rescue us from the brink. If Watson's prophecy stands up, he has exposed a variant of denial that cuts across difference in political affiliation. Will we chase strategies that permit us to avoid radical modifications to our high-energy footprint?

Unquestionably, wizardry is an ingredient to any healthy future. Alternative energy streams are required for our inevitable conversion from fossil fuels, despite the sundry and serious obstacles to their development, scaling, and integration. Still, energy analyst David Fridley argues that while alternative energy streams constitute the spine of any future energy system, that system 'will not be a facsimile of the system we have today based on continuous uninterrupted supply growing to meet whatever demand is placed on it.' The societies we build as we convert from a solid century of dependence on profuse fossil fuels to new energies and new resource-intensive manufacturing processes are yet out-of-focus, but, says Fridley, will 'require us to pay more attention to controls on energy demand to accommodate the limitations of our future energy supply.' In more economically developed countries we do not wonder whether electrons will flow when we throw a switch or liquids when we turn a handle, and this unthinking reliance on vast stores of regularly extractable fossil fuels that are the by-product of millions of years of irregular light from our sun is at the heart of our troubled relationship to energy. Shall we not seek assiduously in this moment to reconstruct individual, corporate, national, and perhaps spiritual relationships to energy? With the dawning of this question we discern a fork in our path that breaks toward denial in its sundry guises or toward a world that looks different from the one in which we were reared and which we are struggling in vain to preserve.

Climate futures will depend on how we engage and speak to one another about climate change and actionable policy. How shall we respond to the persistence of polarising conflict in the face of sufficient, valid, and widely-disseminated science? For Dan Kahan this constitutes the 'science communication paradox,' which is the focus of an emerging 'science of science communication.' Invoking Popper's view that science thrives in societies where truths arise independent of institutional controls, Kahan argues that the simultaneous rise in knowledge and conflict over what is known (the scientific communication paradox) is inevitable but not intractable. In empirical studies, Kahan and his team found that the more scientifically literate and numerate are no more concerned about climate change risk than the least. Rather, they confirmed numerous prior studies finding that people 'conform their assessments of all sorts of evidence to some goal unrelated to accuracy,' and, specifically, that 'assessments of the scientists' expertise depended on the fit between the position attributed to the expert and the position held by most of the subjects' cultural peers.' In brief, beliefs about climate change science, scientific expertise, and the overpowering consensus on climate change are negotiated vis-à-vis the values of the communities to which people belong and with which they identify. Indeed, the most scientifically literate and numerate of opposing cultural groups are more polarised because they run down evidence supportive of their cultural values and bin the rest. Kahan observes that 'when factual issues become entangled in antagonistic cultural meanings that transform positions on them into badges of loyalty,' the logical surmise is that scientific communication needs to 'disentangle' knowledge and identity. Kahan concludes that 'science communication professionals must protect citizens from having to choose between knowing what's known by science and being who they are as members of diverse cultural communities.' He argues, for example, that prior polling about human evolution measures religious identity not scientific literacy, a claim consistent with a 2019 Pew poll finding that unless afforded the opportunity to clarify their views on the relation between God and evolutionary process 'highly religious' folk feel conflicted about the claim that humans evolved over time. In a December 2018 poll of registered voters, the Yale Program on Climate Change Communication found robust bipartisan support for the Green New Deal. 92% of Democrats and 64% of Republicans (57% of whom identified as conservatives) expressed support for policy goals like: 'generate 100% of the nation's electricity from clean, renewable sources within the next ten years.' Republican support for the Green New Deal? Pollsters expressly avoided identifying the party origins of the Green New Deal (the Left) so as to avoid the well-established 'party over policy' effect. They conjecture (in keeping with Kahan's research and the 'disentanglement principle') that while Republicans favour the Green New Deal in principle, they were not aware of its origin on the political Left and were therefore unaffected by partisan loyalty.

Americans increasingly accept and worry about anthropogenic global warming. The acceptance nevertheless trails the science and the worry is not yet scaled to the challenge. 2018 Nobel laureate William Nordhaus argues that US climate policy lags 'miles, miles, miles, behind the science and ... what needs to be done.' Far from closing the interval between our self-destructive here and a distant green there, the 'disastrous policies' of the current administration mean that the US is 'actually going backwards.' Promising empirical research suggests that climate deceit and climate denial are assailable if climate realists and progressives build information environments and employ legal mechanisms that render visible and vulnerable the discursive practice of the climate disinformation complex. Attention to climate communication and the strategic use of disentanglement is integral to mobilising the American public to act on the climate crisis beyond any tribal polarity and every false messiah. Pathways to just and peaceable climate futures are choked by the inducements of capital, a miasma of disinformation, and epidemic minimisation. Negotiating those pathways to the distant green there entails deep change to American dreams of happiness and success, but at a minimum means that next November terminates the aberrant experiment of this truth-indifferent America Made Great Again bullshit.

RUNNING ON EMPTY

Hafeez Burhan Khan

A 5.30pm touch down at Aqaba airport in December 2018. We only have hand luggage so breeze through passport control. There's a bank inside the airport but I can see an ATM just outside. I reach the ATM. 'Out of order', it says. I walk back into customs but the official shoos me away. I tell him the ATM isn't working and I need dinars from the bank. He says I have crossed the line and cannot re-enter. I plead my case to a security guard. He beckons me forward but the official repeats: 'You crossed the line'. He won't let me go. My wife is looking at me with some annoyance. I creep towards the bank gesticulating to everyone and no one. The official grabs my arm and gently pushes me back towards the exit. 'The law is you cannot cross the line'. He says firmly, wagging his finger. Meanwhile our taxi driver has turned up. My wife turns to me and says exasperatedly, 'there are other ATMs in Aqaba'.

It turns out this is true. Our driver takes me to one of the many other ATMs and we then head to Wadi Rum, an other-worldly desert in southern Jordan, made famous by Lawrence of Arabia. It is dark by the time we get to Rum Village. We step out of our taxi and hop onto the back of an open pick-up truck. The truck speeds through the desert valley surrounded by huge black silhouettes of rock. We don't realise how large they are until we see them the next day. The wind is ice cold as it rips through us. I'm having a hoot as we drive across the desert which is lit by thousands of stars. Twenty minutes later we reach our Bedouin camp. Both of us are frozen stiff. We enter the dining tent where there are about a dozen other guests all wrapped up like mountaineers, it is that cold. Food is served, and my wife is shivering too much to even eat. We are shown to our tent. Two single beds. She starts to cry. 'It's so cold', of course it's cold, we're in the desert. We go to inspect the bathroom. Only freezing water is coming out of the taps. My fingers are frozen. I feel sure the real cause of her anxieties is that the toilet and shower facilities are shared. She says she can't rough it anymore. We argue about first-world privileges. Nobody has an en suite in the desert. We are at the mercy of the elements. Welcome to Wadi Rum everyone, welcome to all mod cons and the joy of being at one with nature.

To wake up in the cold isn't much fun. Still, the desert looks spectacular at 8am. Different hues of orange and red battling with the gigantic shadows in retreat. The camp is situated in a bowl of rocks and if you climb to the top, which only takes a few minutes, the view is utterly breathtaking. Huge blocks of black and grey sandstone jut out of the sand like a fist punching towards the sky where the sun is still rising. Across the wadi (dry river valley) and situated at the base of these natural structures are other tourist camps scattered all over the desert. The camp across the wadi from us is a few miles away. The jebels (mountains) are up to 1,800 metres high. What really is staggering is the scale of the place. The camp across the wadi seems just a ten minute walk away. However, we massively underestimate the time it takes to walk anywhere across this vast expanse. Everything is huge, the wadis, the jebels, the aquamarine sky and the sun from which warmth is a godsend. The temperature is only a few degrees above freezing but the intensity of the sun makes it feel much warmer.

Wadi Rum

The other-worldliness of the place is compounded by the fact that we are completely cut off from technology. No phone signal, or WiFi, and electricity is only available for a few hours in the evening. Solar panels fixed on the roof of the dining tent absorb light energy from the sun, so all power-dependent tasks take place after sunset when the batteries are fully charged. It seems such an efficient way to conserve energy, just a pity the solar panels are not able to warm the water for showers.

We walk between epic canyons and after about nine kilometres slightly downhill make out Rum village. There is a bit more traffic with jeeps and camels ferrying tourists back and forth in the distance. We seem to be the only ones walking. Everything is on a huge scale. We stroll alongside a jebel to our right, blocks of rocks that have just shot up vertically. Away to our left the valley widens and slopes upwards to far-off peach-coloured jebels which glisten in the rising sun. The sand is pretty firm which makes walking easy.

Ninety minutes into our hike we arrive at the main junction, and can see Rum Village, an emerging and slightly ramshackle settlement which straddles the width of the valley. In the valley to the right are some enormous sand dunes veering off into the distance. The junction is a couple of miles wide. To our right the valley rears off as far as the eye can see bisecting other jebels. I can see the valley where Lawrence of Arabia's dwelling was, it must have been a good eight kilometres miles away. We begin to veer right to the valley with the sand dunes and as I turn around to look at the jebel that had been on our right, it suddenly dawns on me. Jebel Khazali may mean nothing to you but it's the city on the rock with the Jedi temple, Jedha in *Star Wars Rogue One*. Suddenly, I am giddy with excitement. There is a rock ahead of us, probably 300 feet high with people at the top so we scramble up to get a better view of the Jedi temple. From the top I can also make out the place where the Forest Whittaker character lived. It was such a thrill to survey the numerous canyons calling out *Lawrence of Arabia*, *The Martians* and *Star Wars*. After about twenty minutes on the shoulder of giants, we clamber down and continue on our way.

Crossing this huge sandy expanse, I find this busy route is now frequently being crisscrossed by camel trains and vehicles. About forty minutes later we enter the valley with the dunes and walk up to a strange looking tourist 'camp' which has Martian pods and fairy chimney tents. So this was where the 1% stay. It is even gated with a sandy wall to give the illusion of exclusivity. The wealthy demand all the luxuries of modern life in this most remote of places. The glamping resort even has its own intrusively noisy generator, and I notice a water truck approaching. They must have en suite. How much water must be transported across the desert to service film stars and film crew and other tourists who cannot do without the comforts of home, not to mention the amount of energy expended.

We double back and enter the magnificent Wadi Rum, valley of the moon, itself, eventually leaving the desert to enter a settlement. When I came here in 2002 there were only a few huts. Now a whole village has sprung up. Against the pristine magnificence of the desert, the village is markedly dilapidated with half-constructed, one-storey buildings and plenty of rubbish on the streets. The Bedouin have recently transitioned from a nomadic to a more sedentary existence and I can see that the process of creating an infrastructure to suit a settled area is still in progress. We have our lunch at a clearing where numerous coaches and minibuses drop people off from the airport, before being ushered into 4x4s to sweep them away into the desert.

Jedi Temple

Sustainability is more than just a desert concern for Jordan. A mostly landlocked country, it has very few available energy resources and is struggling to keep up with the demands of its rising population. The good news is that in the last decade the Jordanian government has acknowledged the economy in its present state is not prepared for the challenges climate change will bring and has started to take action, endorsing the National Green Growth Plan, which is aimed at creating a robust green economy.

The plan will focus on agriculture, transport, tourism, waste management, energy and water. Jordan currently imports a staggering 97% of its energy, but by 2025 it hopes that 20% of the energy it uses will be generated within the country from renewable sources. With a climate that enjoys 300 days of sun a year, solar energy is an obvious and cheap source, yet currently woefully under-utilised as testified by our cold showers in the Wadi Rum desert. Only 15% of households contain solar water heating systems. The hope is that this will increase to 30% with increased investment in projects such as the Shams Ma'an Power Plant in southern Jordan, which opened in 2016. The second largest solar farm in the region, it was created at a cost of US$170 million and has 640,000 solar panels, generating 1% of the electricity for the entire country. There are plans to increase that figure to 4%. A new solar park in Maan is in development while other solar projects are being created in Aqaba and Irbid. Solar companies from the United Arab Emirates are also constructing a solar park that upon operation will provide electricity for 42,000 homes and cut 100,000 tons of carbon emissions every year.

Mod con luxury tents, en suite included

Energy is a critical issue, but it is the complex and sensitive issue of water supply that has forced the Jordanian authorities to think creatively in their quest for solutions. According to USAID, Jordan has one of the lowest levels of water availability per capita in the world. The addition of millions of Syrian refugees has placed immense pressure on the nation's resources, particularly as some estimates suggest the population will double by 2047. Since the 1950s, USAID programmes have invested more in water resource management than anything else to improve the water infrastructure of the country. Jordan consumes more water than is available, the renewable water supply only meets a half of the total water consumption and water shortages are caused by unsustainable groundwater extraction including thousands of private wells. Not surprisingly, communities are experiencing tensions over water scarcity especially since the Syrian crisis. The response is to develop more water sources and rehabilitate inefficient water networks.

Serious inefficiencies have been identified and steps taken to address them. These include the fact that 90% of rainfall evaporates or runs off, representing lost potential for capture and aquifer recharge. Agriculture consumes 50% of the water supply yet contributes only 3% towards Jordan's GDP. Around 50% of the water supplied by municipal networks becomes 'unaccountable', meaning it is simply lost, either due to theft, corruption or leaks. In 2000, USAID invested US$700 million in several water networks and waste water treatment plants and a desalination plant by the Dead Sea. It is anticipated that, as a result, clean drinking water will be provided to over one million inhabitants of Amman.

These measures could not be more vital as Jordan grapples with the looming spectre of climate change. According to a report by Stanford University, there will be 30% less rain in 2100 and a 4.5-degrees Celsius rise in temperature in this already rain-impoverished country. There is evidence to suggest that the Syrian war was triggered indirectly by a mismanaged drought. The result is that water flowing to Jordan from Syria has been reduced drastically as the Syrian government now diverts water for its own use.

Acutely aware of the gravity of the situation, the Jordanian government has come up with a pragmatic but unpopular and politically sensitive solution: international cooperation with one of its neighbours. Equipped with the technology and capital necessary, Israel produces 75% of its water from desalination plants and recycles more than half of its waste water for agricultural use. The Red Sea/Dead Sea Project is a US$900 million agreement between Jordan and Israel to desalinate Red Sea water, producing 80 million cubic metres of clean water annually in a desalination plant in Aqaba. The by-product, brine, will then be sold to southern Israel, and pumped into the Dead Sea to raise water levels. Consequently, Israel will release 50 million cubic metres of water from the Sea of Galillee to a reservoir in northern Jordan and Palestine. Israel's motives in pursuing this project are strategic. It has a vested interest in the stability of Jordan set against the region's instability.

However, the backdrop of the Palestine-Israel conflict inevitably impacts the progress of the initiative and most recently the shooting of a Jordanian teenager by a member of Israeli embassy staff in the summer of 2018 caused the project to be suspended. Further tensions have arisen due to a water sharing deal signed by both countries in 1994, which Jordan says is unfair because of the dwindling flow of its rivers and the explosion of wells in Syria which have drained the rivers of their water. King Abdullah II bin Al-Hussein has threatened to pull out of a deal where Israel has been leasing agricultural land when the lease expires this year. Prime Minister Benjamin Netanyahu has promised alarmed rural communities in Israel that an extension to the deal will be negotiated but the king faces wider public pressure to walk away from the much derided agreement.

The Yarmouk River in northern Jordan has been reduced to a trickle. Some reservoirs are running at 20% capacity due to erratic winter rains, while the River Jordan is a fraction of what it was when it enters the Dead Sea. Only fifty years ago, the Dead Sea stretched fifty miles north to south, today it is thirty miles and the sea level is falling at a staggering rate of 1 metre per year. In 2010, Jordan had a population of 7 million, four years later the population had shot up to 9 million largely due to the flow of refugees from Syria and Iraq. This year, the population has topped 10 million, a 40% increase within a decade. Any industrialised country would find it hard to cope with such an increase.

The cost of building a desalination plant without Israel would be prohibitively expensive. The 320 kilometre distance to transport the water from Aqaba to Amman would cost billions of dollars, so there is an awful lot at stake. Last year the king was left with no choice but to sack the prime minister due to the abolition of bread subsidies which led to massive protests all around the country. The fear is that growing water shortages will provoke even greater and more violent unrest. Presently, residents in refugee camps only have access to water once a week. In rural areas it can be as little as once every three weeks. Most buildings contain water storage containers, so families may store the resources they do receive. Yana Abu Taleb of Eco Peace in Jordan, an organisation focused on environmental peacemaking in the Middle East, notes that 'on average Jordanians receive 45 cubic metres of water per year, half of what they were [receiving] before the Syrian crisis'. To put that in context, this figure represents only about two per cent of the amount of water Americans consume on average each year. According to Samer Talozi, a professor of science and technology and director of Mirra think tank on irrigation and agriculture, the Azraq aquifer is currently being pumped at double its safe capacity. Of the country's twelve main aquifers, he believes ten are now massively depleted. People are now digging wells that are ten times as deep as they were in the past to try to access the water. In the ancient city of Jerash, eighty kilometres north of Amman, which was once littered with Unesco-listed Greco-Roman ruins boasting lavish springs and baths, desperate farmers are going bankrupt and being forced to sell off their land because the water crisis has ruined their crops and livestock.

Jordan's attempts to address its sustainability problems are not all rooted in futuristic technology. Inspiration has also been derived by reverting to the past. In the town of Umm el-Jimal, located by the Syrian border, locals are restoring reservoirs built nearly 2,000 years ago, so they can once again be utilised for water retention. For nearly 800 years, winter rains and run-off from the mountains in Syria were streamed through canals into the basalt reservoirs, which stored the water through the dry season. In 2015, the first of such reservoirs, the size of four Olympic-sized swimming pools, was restored. Through the work of the Umm el-Jimal Archaeology Project from 2012–2019, in conjunction with archaeologists from Calvin College, in Grand Rapids, Michigan, the community is now able to reuse the developed and reactivated water system. Traditional and contemporary methods have enabled the construction of a 3.5 kilometre water supply channel to the reservoirs, alleviating water shortages that had been previously experienced.

Research into water sustainability is occurring in institutions and centres across the world and Jordan is ready to explore all the options. The country has been particularly successful in forging international cooperation by attracting private finance from abroad, largely due to its geopolitical importance, relative stability and public investment by the government. In

February 2019, the then British prime minister, Theresa May and King Abdullah, co-hosted 'The London Initiative', which was 'designed to lay the foundations to unlock growth, jobs and investment for Jordan. The initiative marks the start of a new partnership approach between Jordan and the international community in pursuit of Jordan's sustainable growth and self-reliance.' The conference brought together organisations from over sixty countries. According to Talozi, Jordan is an international test case: 'If we can build systems that work in Jordan, they will work everywhere.' A host of research institutions are now working to do just that. Massachusetts Institute of Technology (MIT) has been testing low water pressure irrigation technology, which they hope will need less water and energy and will use solar power. Stanford University has been developing software which uses external factors such as urban growth and water prices to guide decisions about replacing or repairing water infrastructure and by placing facilities so they do not affect the groundwater. The Helmholtz Centre for Environmental Research, Leipzig, has been testing soil filtered waste treatment facilities which lessen leakage in large plants, preventing water pollution.

As for the Red Sea/Dead Sea Project, Israel signalled its eagerness to restart the project pledging US$1 billion – recognition of the strategic importance of the stability of Jordanian security and rule. Both Saudi Arabia and Israel wish to limit the threat of ISIS or Iran so geopolitical considerations are imperative in their desire to invest in the nation. There is a caveat for Jordan, however. It is speculated that Jordan would be compelled to scale back its support for Palestinian self-rule in exchange for water security and surrender custodianship of Al Aqsa Mosque in Jerusalem to the House of Saud. For King Abdullah, all the options are loaded. The quest for sustainability is irrevocably politicised in a region fraught with competing interests. There will be winners and there will be losers, lines will be crossed and there can be no going back. I hope we can reflect on that, and contemplate quite what is at stake next time we find ourselves in the desert without an en suite.

REFUGEES

Tawseef Khan

Think of the word 'refugee'. What kind of image surfaces in your mind? Personally, I start thinking about the photographs of refugees scrambling into already-packed trains through the windows, anxious to leave Croatia and get to Western Europe before the borders closed. I think about the countless pictures of refugees packed into tiny, swaying dinghies like sardines in a can, then landing on the islands of Lesbos, Chios, Samos, for example, and being wrapped in gold and silver foil blankets. Those images, like the more recent images of the human caravan travelling through Central America towards the US, depict desperation, despair and chaos, but also the sheer will of humans to survive.

My mind turns to starker images of the current 'Refugee Crisis'. The image of three-year-old, Aylan Kurdi's body, washed up on a beach in September 2015 in Turkey. Or the figure of five-year-old Omran, sat expressionless in an orange chair, caked in dust and dried blood, providing a human face to the suffering of Aleppo. These pictures have already come to symbolise the senselessness of war, its horror, the innocence of many victims, and the indiscriminate way in which they are chosen.

And when I think of refugees, I think of the way some pictures that illustrate how we – those of us who are receiving refugees – have come to dehumanise them too. In 2015, a Hungarian journalist was caught tripping over a Syrian man trying to run from the police whilst carrying his child. It caused uproar within the international community. The lines of refugees coiling across a field, waiting to cross the borders into Western Europe was famous too. It was slapped with the words 'Breaking Point' and unveiled as a campaign poster for those wanting the United Kingdom to leave the European Union (EU). We will look back on these images one day, and find that they define our times; they speak so profoundly about the condition of our societies.

And, now, when I think of refugees, a new image emerges in my mind's eye. On 20 October 2009, the Maldives government held its cabinet meeting some six metres below sea level. Ministers, in their scuba gear, communicated using hand signals and white boards, and signed a document urging all countries to cut their carbon emissions. Climate change, they announced, is threatening the very existence of the Indian Ocean island; and may force its 350,000 inhabitants to become a new type of refugee – those escaping the adverse effects of climate change.

When I think about these pictures, though I cannot identify with the specific experiences, I do, nonetheless, identify with the refugee story. I grew up listening to the tales that my mother and grandmother told, of relatives in Uganda fleeing the Idi Amin regime of the 1970s, of people dying and being born in the makeshift refugee camps set up around Heathrow Airport in London. There is value in remembering how, even in our recent history, the 'refugee story' has not only played out in some distant region, but also nearer to home – on British soil, against a group of people that held British citizenship. During this time, 27,000 Ugandan-Asians who had British citizenship ended up fleeing to Britain. When I was growing up, refugees, their plights and their futures were not distant at all, but tied to my own past, present and future.

My father is a lawyer. When I was a child, he chose immigration law as his specialism and our living room as his office. I was only four years old at the time; listening to clients rattle the brass door-knocker, the door swinging open as they stepped into our living room. I overheard conversations about asylum interviews, appeal hearings and the Home Office, and in the evenings, when our house was quiet and empty, I would slip round my father's desk and sit on his chair. I would grab his telephone and hold it to my ear. Scrawling on the notepad, I would advise my imaginary clients. I would write down their narratives of persecution. I would arrange their interviews with the Home Office. I would tell them to stay strong. My parents have the photographs.

It was no great surprise, then, that I followed my father into the family profession – immigration law. And it was no surprise either when I decided to pursue doctoral research on the refugee experience. Most of my life has been devoted to thinking about refugees and religion. And I have come to the conclusion that, for those who identify with Islam, the refugee story belongs to us all. The early inception of Islam is tied up with the experiences of persecution and safe refuge. Indeed, for many of us, the term, 'refugee' will bring to mind the *hijra*, when the earliest Muslims fled the mistreatment of Macca for the safety of Madinah. But my point is not just that we are familiar with narratives of persecution, but that, perhaps because of this experience, Islam is especially clear on the Muslim duty towards refugees. Indeed, as a number of Muslim scholars have argued, asylum is an integral part of Islam's approach to human rights. The connection between Islam and asylum makes sense. Islam is intrinsically inclusive, and it embraces difference of all kinds. We can think of this in terms of the differences that force people to abandon their homes – race, religion, nationality, political opinion, gender, sexuality – but we can also think of difference in terms of immigration status; those of us who are documented embracing and assisting the undocumented.

So what does it mean to be a refugee at this moment in time, and what might it mean to be one in the future. It is impossible to imagine a future for refugees without first thinking about their numbers and the conditions that create them. The figures are breathtaking. For the last six years, the global refugee population has consistently grown; it is the biggest it has ever been since the Second World War. According to the UNHCR, by the end of 2017, the number of displaced people in the world stood at 68.5 million – an increase of 2.9 million on 2016. Of this 68.5 million, 16.2 million were displaced during 2017. It amounts to one person being displaced every two seconds, or 44,500 people per day. One in every 110 individuals in the world is displaced. It is equivalent to the population of Thailand.

The vast majority of refugees are a product of conflict and war. This comes as no great surprise, given the detailed and often graphic depictions of war splashed across our media. The notion of conflict and its victims conjures up some of the images I set out above; the wars in Syria and Yemen. We may think of the war in South Sudan, the crisis in the Democratic Republic of Congo, the way that the Rohingya have been shut out of the Burmese national myth and their persecution endorsed by the state. Two thirds of the world's refugees come from just five countries – Syria, Afghanistan, South Sudan, Myanmar and Somalia – countries that have all been marred by prolonged internal conflict.

It is no leap of the imagination to suggest that, for the foreseeable future, conflict will continue to be the primary cause of forced displacement. War feels like it is everywhere, and if it isn't, fighting talk certainly is. Our states are increasingly patrolled by ignorant and bullish leaders. Our societies seem more and more divided along different sectarian lines. It can often seem like we are sliding inevitably towards a perpetual state of rivalry and conflict.

As much as the world's conflicts have captured our attention, it would be shortsighted to focus on them as the only source of refugees. The greatest threat to our planet comes from climate change, not war. And it is climate change that will, in many ways, dictate the future of refugees.

According to a report published by the World Bank in March 2018, the effects of climate change on people's movement will be visible within the next ten years. In its study of the populations of Latin America, South Asia and sub-Saharan Africa, the World Bank predicted that by 2050 climate change will have forced approximately 143 million people to relocate within their national borders. That equates to 3.5% of the combined population of these regions. Shocking as this may be, it is actually a conservative assessment. It does not include those affected by short-term climate 'shocks' like floods, cyclones and droughts. It does not consider those who seek refuge outside of the national borders. It provides no insight into the impact of climate change on the rest of the world. For an idea of the scale of this potential crisis, the International Organisation on Migration (IOM) estimates that by 2050, there will be 200 million people displaced by climate change across the world.

In December 2017, *Science* magazine published a study which argued that the numbers of refugees entering the EU will dwarf anything we have experienced this generation. Even if, for example, the pace of climate change was stabilised by 2100, the number of asylum applications would increase on current figures by 28%. If global temperatures grew at their current pace, asylum applications would rise by 200%. A million refugees would enter the EU every single year. The figures are incredibly sobering.

In all of this, the Muslim world is especially vulnerable to climate change. The vast majority of the world's 1.6 billion Muslims live in large, dense populations within fragile eco-systems. Countries like Morocco, Ethiopia and Iran are either already drought-prone or now especially susceptible to drought, whereas Pakistan, Indonesia and Malawi are at risk of flooding. As global temperatures rise, again, the impact upon the Muslim world would be especially acute, with summer temperatures rising at twice the global average in the Middle East and North Africa region and 1.5 times in the West African Sahel region. By 2025, Muslim world will face an acute shortage of freshwater, which is already an issue, given the fact that 1.4 billion people in the developing world have no access to clean and safe drinking water in the first place. There are plenty of solutions to the challenges posed by climate change (Saudi Arabia and Morocco have started to build solar plants), but given the prevalence of other problems in the Muslim world, especially of a political and economic nature, the problem of climate change is still not being given the urgent attention that it deserves.

Bangladesh is an especially frightening case study of how climate could affect Muslims around the world. Although the country contributes only a small fraction to the world's carbon emissions – a mere 0.4 metric tons – it is suffering already and will suffer far more than the world's biggest polluters. Bangladesh is mostly flat, sat at the foot of Himalayas and held together by rivers and deltas. In just thirty years, temperatures in Bangladesh will increase by 2 degrees Celsius. In sixty years, sea levels will be two foot higher. Himalayan glaciers will melt; cyclones shall batter the coast. Agricultural land will be ravaged by seawater and soil erosion. Drinking water will be contaminated. In Bangladesh, the impact of climate change is already being felt. Of the 400,000 people that move annually to the capital city, Dhaka, the International Organization for Migration (IOM) found that 70% moved after an environmental disaster. Climate change is already uprooting people from their homes, but our leaders sit on their hands and do little.

As the government of the Maldives highlighted in the famous image of underwater cabinet meeting, it is at the frontline of climate change. Forged out of a collection of 1,200 islands in the Indian Ocean, 80% of it sits at less than a metre above sea level. Even by the most conservative estimates, rises in sea level could wipe out the islands completely. As soon as 2030–2060, in fact, scientists expect islands like the Maldives to slowly become inhabitable – almost a century earlier than first expected.

Faced with the threat of outright extinction the Maldivian government announced plans in 2008 to purchase land elsewhere for the population to relocate to. Since then, however, government strategy has changed course. The government has chosen to use wealth gained from tourism and leasing islands to wealthy nations like Saudi Arabia, to fortify the most vulnerable and populous islands, and even build new ones that could adapt to rising sea levels. One such purpose-built island could accommodate up to 130,000 people. Eight such islands have been completed, one will be completed in 2023, and three more are planned. Given the 400,000 population of the Maldives, this is impressive, but not quite enough to take them out of the danger-zone.

For the Maldives to survive, the coral reefs that the islands are formed from and surrounded by must survive too. The reefs are essential; the tourism and fishing industries are dependent on them, and they buffer the shoreline from heavy waves and erosion. As sea levels rise, Maldivian freshwater supplies will be contaminated, and it is heartening that the government, through US aid, are building water treatment centres. But rising sea temperatures compound the dilemma; if the coral reefs die, so too will much of the Maldivian life that is constructed upon it.

And yet, with all this said, climate change may not just make parts of the world uninhabitable. It could also create conflict. That seems to be the contention of a 2015 study into the causes of the Syrian civil war. The researchers argued that the drought which gripped Syria and Jordan between 2007 and 2010 displaced Syrian farmers, triggering communal unrest and eventual conflict. Although other researchers have disputed this argument, as temperatures rise and larger parcels of agricultural land are lost, it seems only logical that climate change could sew chaos and discord.

One of the clearest examples of a climate-induced conflict being averted was between Egypt and Ethiopia. Egypt, which gets all of its water supply from the Nile, was so alarmed by Ethiopian plans to build a hydroelectric dam on the river that it threatened all-out war. Through the second half of 2018, the two countries came together to hash out an agreement, with Sudan, where all three parties reliant on water from the Nile agreed not to restrict each other's access to the water supply. But as it stands, the detail of how these rights will be secured for all three countries is unclear.

The connection between climate change and conflict seems obvious when we look to the countries most susceptible to climate change. Pakistan, Bangladesh, the Philippines, Oman, Sri Lanka, Colombia, Mexico, Kenya and South Africa – the ten most vulnerable – also happen to be developing countries at the nexus of existing political and economic challenges. Some already host significant refugee populations. Most of them are amongst the least equipped to deal with climate change. By contrast, the five countries least vulnerable to climate change – Finland, Sweden, Norway and New Zealand – are notable for the fact that they are some of the richest, most peaceful and politically stable countries on earth. They will have no issues dealing with the consequences of global warming.

The logical response is for global leaders to commit to limiting the pace of global warming – and fast. Yet again, we need the richest nations to take the lead, and we need the developing world to be the focus. As Oxfam has revealed, the world's poorest 50% are responsible for 10% of global carbon emissions. By contrast, the world's richest 10% are responsible for 50%. With the weight of political power also lying with the world's richest nations, they must take the lead in prioritising the environment over perpetual economic growth. But critically, the poorest nations also need to do more on climate change. As their economies grow, they should reduce their reliance on fossil fuels in favour of greener alternatives. They should invest in alternate technologies and agricultural innovation to combat the loss of agricultural land. They should prepare for the strain that rising temperatures will inevitably bring.

To this end, the Paris Agreement, which was adopted in 2015 by 184 countries, gave us hope. It united most of the world on a course of action to limit the rise of global temperatures. But with President Donald Trump announcing in September 2017 that the USA would leave the deal, translating the Paris Agreement's broad language into a concrete set of rules has become even more difficult. With the USA gone, a number of developing countries are trying to create separate rules for rich and poor countries, and this threatens the deal altogether. Moreover, with the agreement focusing on voluntary targets that are set by the nations themselves, rather than mutually agreed, legally binding ones, we are still sleepwalking towards disaster. Scientists consistently point out, for us to succeed at curbing climate change, the agreement needs to demand more of its signatories. If it does not, we will begin to see the fallout in the creation of an unprecedented number of refugees.

The human ability to survive through adversity is what characterises us. Our ability to find difficult and potentially lasting solutions to problems through a mixture of intelligence, tenacity and sheer will. It is this ability that shall dictate the future of refugees. Whether we can find pragmatic, perhaps radical solutions to extremely complex problems. Whether we are sufficiently survival-minded to prioritise human life and a healthy, sustainable planet over infinite economic growth. I foresee the future of refugees resting on these skills. I see these challenges coming together over three overarching questions:

First, can an increasingly populist, isolationist and (let me just say it) selfish public recognise the interconnected nature of our world, to a degree that would enable us to give the lives of refugees the respect and value they deserve?

Secondly, can our global leaders begin to act with the statesmanship required, using diplomacy to resolve conflict, allocating funds where needed, committing to finding solutions to arduous, but pressing global concerns?

And finally, can we create the necessary policies and instruments – and stick to them – so that we can support refugees properly when they are created, and reduce the numbers that are forcibly displaced and remain displaced for generations?

As we think about the future of refugees, we may have to rethink how we define the term itself. After all, climate change is not a recognised form of 'persecution' in law. Moreover, the reasons for forced displacement are no longer as simple as conflict and persecution. For the countless cases of the individuals driven to seek a better life in the West, poverty is regularly entwined with persecution. There's an argument to be made that poverty is a form of persecution too. What remains clear, however, is that the future of refugees requires us to change our current ways of thinking and responding, including how we create and conceive of refugees as a group of people.

PROSPEROUS COMMUNITIES

Shanka Mesa Siverio

Shocked by my own bulbous carbon footprint – which I recently calculated on the misguided expectation that I would be well below average – and inspired by the recent wave of climate change protests around the globe, I decided to take some climate action from a very local perspective. It turns out that my habit of flying places adds significantly more to my carbon footprint than pretty much everything else I spend my time doing, so local seemed logical, if only to keep myself busy within walking distance of my house. Despite the long history of failed local groups I have been a part of, I have a soft spot for community-led action since it appeals to my (perhaps naïve) belief that cooperation and goodwill trump individualism in the long run and that community life is good for the soul. So, with a sense of hope in the power of the neighbourhood and a certain personal responsibility for the state of the nation's carbon emissions, I attended a small group meeting for a Wednesday night brainstorming session on ways to mitigate the climate crisis from within our area. I left the meeting with less personal guilt, but with more frustration and a feeling of discord. On the face of it, nothing particularly unusual had occurred and some of the discussion had been both informative and actionable. So why my unexpected unease?

After a little reflection it wasn't difficult to identify the aspects of the dialogue that had prompted less concern for my air miles and more about the social divisions which often get exposed at such small community gatherings. Amidst the general positivity and convivial atmosphere I detected a subtle but palpable undercurrent of judgement, resentment and emphasis on individual action that felt out of proportion with the immense scale of the climate crisis. But it was not just this that I found jarring, it was the focus of comments and who they were aimed at that prompted my feelings of defensiveness. It was particularly hard to understand the relevance of fried chicken to the debate and though I am no great fan of plastic bottles or bags, I couldn't help but note that these and other offences that were singled out had less to do with carbon mitigation and more to do with class-preferences and recent movements towards certain 'green' lifestyle choices. As the only non-white member of the group and as someone who is more than averagely preoccupied by urban social issues, I found it even harder to miss the socio-economic and racial disparities that this discussion was highlighting from within our middle-class enclave. After some further consideration of the scene that evening I came to the conclusion that many of the lines were borrowed from the scripts of 'conscious consumption' movements that are currently permeating into mainstream culture.

Prompted by the larger narrative on climate change, such movements are busy convincing us that climate change is an issue that must be addressed primarily at an individual level, by choosing to buy the right things over the wrong ones and ensuring others do the same. They rely on persuasion and have a distinctive aesthetic quality to them, embodied by the new wave of 'green lifestyle influencers' on social media. These influencers, whilst highlighting the very real need to be less wasteful and more environmentally conscious, are also using the message to gain followers to promote various lifestyle choices that include plastic-free, zero-waste, veganism and minimalism. There are notable visual and cultural tropes being sold alongside bamboo toothbrushes and organic produce in the process. Both directly and indirectly, these associated images are conflated with environmental moral superiority and are not hard to miss. This is not particularly new if we consider that within most marketing strategies, visual advertising and product placement are used to suggest that certain products and consumer choices will lead to an elevated social status and belonging. The key difference now is perhaps the emphasis on moral superiority and the environment. In this new age of environmental awareness, the symbols of success are at once different but strangely familiar. They still include what vehicle you drive (electric or even better, a bamboo framed bike), the brand and type of your hand bags and shoes (vegan leather or recycled plastic bottles ideally), where you go on holiday (ideally a staycation in an off-grid yurt) and the general contents of your canvas shopping bags. Much of the aesthetic appeal of these iconic images spring from aesthetics traditionally associated with morality, spirituality and purity in the West. Specifically, clean, bright, youthful, white and predominantly sold by young female influencers who often have the benefit of hailing from dominant versions of beauty and culture.

Whilst many of the followers of such movements would deny that they encourage judgement or exclusivity, there certainly appears to be a stark picture of what cultural and moral supremacy looks like. Though not compelled to do so, we may be quick to judge those who appear to conform less to such ideals as morally and culturally inferior, if only subconsciously. Eating fried snacks and using plastic bags are easily associated with tribes that are then labelled as not caring about the environment. That the shop keeper being judged, prior to a conversation or invitation, for use of plastic packaging is Bangladeshi and the teenagers under verbal attack for frequenting fried chicken shops are African-Caribbean is less likely a coincidence and more likely a reflection of the pre-existing biases now fuelled by marketing, and a political agenda that seeks to ensure just enough division within society for stability. These biases are often socio-economic at their core, with race as a bi-product, and ignore the real barriers faced by lower income groups in such cultural norms as recycling and 'ethical consumption'. Some of these barriers have been highlighted in a study by the Economic and Social Research Council on Environmental issues and Human Behaviour in Low-Income areas in the UK. The report concluded that whilst there was little to suggest that those on lower incomes are less concerned about wider environmental concerns than others, they face different barriers to action and are incentivised differently to more affluent groups. Unfortunately, to the average unrelated observer without such studies or personal experience of difference to hand, it is easy to dismiss whole groups based on the messages we now receive from companies and influencers with a 'green' agenda.

Adding to the complexity of the community level response to big issues are the common limitations of being human in the twenty-first century. Firstly, the ability to unite against someone can often be a psychological advantage in holding together social groups. In psychology this is called the 'common enemy effect' and if you've ever been part of any group in which cooperation and bonding are required, you will likely have experienced this in action. Secondly, logistics. Rather than address the structural problems that exist where climate change is concerned, at community level, it can be tempting to punch down, to the individual level. Unfortunately for society at large, our anger is simply more easily directed at the local 'villains' who do not conform to our mental picture of 'green' than it is at the FTSE 500

CEO who is willing to sacrifice the earth for a steady 10% profit margin or the politician who fails once again to sign up to the necessary policies for change for reasons of access alone. Finally, there is the inevitable problem of what brings these groups together in the first place: urban geography. Like most communities of geography, we were a group reflecting the white, middle-class bubble in which we live, despite being surrounded by several more deprived areas with large minority ethnic populations. This is not the first time I have been the only non-white member of a community group, nor will it be the last until actual barriers to diverse participation are addressed in a meaningful way by those with the power to do so. Ultimately the seemingly random make-up of our group was determined by the market driven economy that dictates where we can afford to live and who we socialise with.

The influence of the free market has an even more directly discernible impact on climate justice if we are willing to look hard enough. The recent shift towards conscious consumption has brought to light many of the potential pitfalls of allowing the agenda to be determined by profit seeking organisations. It has already led to such things as 'green washing', whereby products and services are sold as 'good for the environment' whilst the reality is hidden by jargon and complex practices such as off-setting. The phenomenon of off-setting is one of the activities that leads to uncomfortable questions around the resurrection of 'green colonialism'. Commonly practised in the colonial era (and after) as a way of obtaining and disposing of materials and waste to fuel industrial development without cost to home environments, in today's context, environmental colonialism is taking new and more inventive forms. It now includes various activities that allow developed nations and corporations to position themselves as leaders in the race to cut emissions, whilst creating social and environmental degradation in developing ones. An illustrative example includes forced evictions and food shortage caused by the purchasing of agricultural land for forestation for carbon off-setting in areas of Uganda, Mozambique and Tanzania by Norwegian companies, as brought to light in research by the Oakland Institute. In similarity to the environmental colonialism of the past, the mechanisms by which this and other injustices occur in the name of ecological progress rely on inequality and the existing power structures. Significantly these have not changed since the colonial era. The surge of such practices can be attributed to the continuing search by private companies for growth and profit on a finite planet, whilst otherwise 'sticking to the rules' imposed by governments with unreconciled ideals of economic growth and environmental preservation. Such social and environmental failings by individual companies are essentially incentivised by the neo-liberal economies and their associated financial systems, which continue to invest in areas which can provide the primary goal of rapid growth (including many carbon-heavy commodities and services) and regulate only up to a point. Moreover, we are all complicit and reliant on these practices as our own pensions and savings are hopelessly bound up in this financial system and bank on this continued investment growth for some of our most basic needs. In a society that hinges on growth for stability in the age of climate crisis, there seems to be only one option available: green growth.

However, many green growth models make several assumptions that require a degree of serious optimism or for us to limit our ambitions to within our national borders, ignoring effects elsewhere. Under all green growth models, citizens are still responsible for consumption at a rate that cannot be sustained without the total decoupling of fossil fuels and consumption, facilitated by technology. Not only does this leave the existential and social problems that a consumer society creates unaddressed, at present we do not possess the technology to achieve this radical solution. Given the predicted timescales suggested by researchers such as Chatham House think tank, the deployment of such technologies would take between nineteen and thirty years to achieve. This leaves around ten for the free market, through competition between private interests and innovation in the quest for profit, to develop such technologies. This short time-frame will determine whether profit is enough of a motivator to the world's inventors. Whilst there are several companies who have recognised that green technologies are the future, the current lack of regulation and continued investment in fossil fuels means that the chances of innovating the necessary technologies are limited by funding, resources and even a lack of competition which is often touted as the primary driver of invention in the free market. As summarised in a recent report on Green Growth by LSE, 'the social returns exceed the private returns, perhaps on average by a factor of four, so the private incentive to innovate is less than is socially desirable'. These are not good odds.

Whether you choose to remain optimistic about the free market and technology or not, economic and ecological arguments are also philosophical and moral. As suggested by economist Tim Jackson 'our ecological debts are as unstable as our financial debts. Neither is properly accounted for in the relentless pursuit of consumption growth'. Up until very recently, it has been possible for us to turn a blind eye to these double debts as they have been felt most strongly by the poor as inequality in the West has increased and the impacts of climate change have been primarily felt in the Global South. As if to demonstrate the principle that life isn't fair, the debt is currently being paid by those who have contributed least to the problem; by societies who are the lowest emitters of greenhouse gases and whose resources were in many cases exploited to service growth elsewhere throughout the colonial and neo-colonial eras. In turn, this historical growth is embedded into our society in the physical infrastructure, institutions and urban fabrics and is exactly what makes us more resilient in the face of climatic variations. The fact that much of the East and Global South now aspires to the model of development pursued in the West should be of no great surprise. Yet just as it appears to be within reach for some developing nations, many otherwise fair-minded individuals balk at the very suggestion on the grounds of climate change. Global policy, though mixed, seems to concur with an expectation that the status quo must, at least in some sense, remain intact and that the historical advantage should not be a factor in future pledges.

A possible alternative to this trajectory is an idea that is built on the simple premise that individual prosperity relies on the prosperity of the whole; that viable alternatives to endless economic growth do exist. This was the central idea behind the 2009 report published by the Sustainable Development Commission, *Prosperity Without Growth*. The report argues that it is both possible and desirable for developed nations to achieve a steady state economy, whilst allowing poorer countries to catch up and eventually reach a steady state at a point at which enough wealth is embedded in their systems, as well. Pivotal to the author's arguments is the redefinition on what we have come to understand as the good life: 'prosperity consists in our ability to flourish as human beings – within the ecological limits of a finite planet'. The idea behind this is that whilst our basic needs must be met via economic growth, many of our other needs such as a need to belong, to participate in society in a meaningful way and to have a stable life within which to nurture families, relationships and social bonds are not able to be met by, and are often in conflict, with the endless growth model, which by its very nature requires us to constantly compete and to trade much of our time and energy for consumer items and experiences that neither bring us fulfilment nor can be sustained by the limits of the planet. The report was published ten years ago in response to the 2008 economic crash and challenged the current economic model on social, economic and environmental grounds. Given the state of UK politics and society which sees increasing division and racism, according to recent statistics, it is perhaps telling that much of the opposite of what is advised in the report has been implemented during this time. It is also unsurprising that the average carbon footprint for a UK citizen is about double that of an average citizen of the world.

The observation that nature cannot be ignored within any economic model for the future is now easy to acknowledge except by the few die-hard climate change deniers or those hoping to escape to another planet, as we have increasing first-hand experience of the ecological effects of over-consumption. Planetary boundaries alongside human needs literally shape economists Kate Raworth's more recent popular model, 'doughnut economics'. This model also makes some quite logical conclusions for the economy and we are asked, in the face of new evidence, to acknowledge that whilst growth is indeed needed up to a certain level of development, endless growth is neither possible nor desirable for quality of life overall. Beyond a certain point, which most developed nations have passed, a more cyclical economy is proposed. Such ideas are not entirely new and have been put forward by many separate economists at different times. Stuart Mills, for example, based his nineteenth century theories of growth tapering off into a steady state economy on Christian views on the 'art of living' and nature and projected limitations of agricultural yield. More recent theorists have challenged long held notions of what drives economic decisions from a perspective that looks back on the last century of industrial growth with the benefit of hindsight. In all of these steady-state or de-growth models, the philosophical understanding of both nature and people differs fundamentally from those that inform our current neo-liberal endless growth models. Aside from the mass of climatic and ecological data now informing the debate, the growing science and psychology of happiness and well-being provides the basis for many of the arguments put forward by the supporters of steady-state economics. Concepts such as Gross National Happiness (GNH), which uses alternative indicators to measure global progress have had a huge impact on current thinking. GNH was originally developed by the Buddhist nation of Bhutan in the 1970s and pioneered a quantifiable approach to quality of life and environmental standards in a way that married Eastern philosophy with Western mainstream economic science by challenging the use of GDP as a measure of human progress. The measure has since opened up the conversation to allow us to question the models of development and progress the world aspires to and challenge the dominance of economic growth as a policy driver.

At the heart of both steady state economic theory and many new measures of human progress is the core belief that human well-being is not at odds with natural systems, rather the two are intrinsically linked and therefore must be considered holistically. As we reach a crisis point with the effects of climate collapse reaching ever closer to home the need for self-preservation, if not the moral imperative, compels us to redefine our current relationship with nature. It is perhaps this recognition that has seen traction for 'de-growth' increase. The de-growth movement, with input from steady-state economists and philosophers, questions the relationship between consumption, marketing and production, suggesting that a simpler life with both less production and consumption, but more time for leisure and community may be more rewarding from a human perspective and more sustainable from both an economic and environmental one. It requires change at both a local and global level and hinges on a changing relationship to consumption and labour and not just technological fixes. It is also a call to put human progress and well-being at the heart of what is now commonly termed post-growth economics.

These concepts have implications that go beyond the practical and prompt us all to question deeply held notions of the good life and our own world views. In acknowledgement of the important role that faith-based communities have to play in bringing about the implementation of such ideas, Odeh Al Jayyousi in his work with the UN, has argued that Islam can offer a unique interpretation of de-growth principles of relevance to Muslim communities. Since Muslim-majority countries are amongst those with both the most to gain and lose in the transition towards a non-carbon economy and that Islam is one of the most commonly practiced religions, this could have profound importance in the direction the world now takes. This is not to suggest that the deciding factor is faith; we have plenty to suggest it is not. Take for example the fact that the oil-rich Middle-Eastern countries with extremes of temperature as well as wealth are amongst the worst polluters, whilst Morocco and the Gambia are the only two nations with sufficient policy pledges to operate and stay below the 1.5 degree warming target despite their shared religious roots. Clearly the socio-political and geographic differences between these best performing and the worst performing are complex and well beyond the scope of simplistic observations on religion and culture. However, the possibility that communities of interest and faith might have as much to offer as communities of geography and politics is a hypothesis worth exploring for anyone looking for real change in the way we respond to the challenges ahead.

So, what might an Islamic de-growth agenda look like in practice? Al Jayyousi has evoked concepts of balance (*mizan*) and proportion (*mikdar*) in relation to the natural world, as well as the concept of living lightly on the earth (*zohd*). Subjective as these concepts are, when viewed in light of environmental data alongside per capita emissions and consumption, a sense of balance suggests that it is at the global scale of the balance that must be achieved, even if the concepts should be interpreted locally within this global context. Echoing this, the Islamic concept of *tawheed* or 'unity' has been interpreted by Al Jayyousi as 'a unified mission of humanity to protect the Earth or construction of the Earth'. Inevitably, any such global mission must accept the extremely imbalanced nature of development to date and perhaps seek to redress this in some ways for the sake of unity. Like de-growthers, this view accepts that action need not be even, and that growth may still be a requirement for poorer nations. Another, perhaps more important interpretation of such 'unity' is the implicit call for national boundaries to be bridged. How countries of extremely different cultures and geo-political contexts might seek to come together in a shared mission remains another question.

De-growth principles suggest that there is much to be learnt by the West from developing countries, traditional societies and the non-dominant cultures that have practiced ways of life that have either been squeezed out of mainstream consumer culture or have never held much sway. For example, steady-state economist Jackson has noted the similarities between much of the theoretical perspectives in his work on prosperity without growth and the concept of Ubuntu, the African concept of shared humanity in an explicit recognition that there are other relevant forms of knowledge worth pursuing in the quest for solutions. Research using GNH indicators has made significant progress towards persuading western practitioners to accept that dominant models of progress and development are not necessarily dominant for the right reasons. Such thinking is the very opposite of what we see in many of the green consumption movements which seek to promote consumer culture and capitalist ideals in greener ways, ignoring the ancient and often non-Western roots of some practices that are now being re-packaged and re-sold for markets finally ready to buy into them. Under de-growth principles, such ideas need not be sold so overtly as policy would ensure that most are freely available and require only a shift away from status driven consumption, towards shared economies developed locally, reduced hours in paid work in favour of time spent with family, friends and community or creating a slower pace of life more connected with nature; in short what has been the norm for much of the economically middle-class outside of the developed West.

Of course, this takes an idealistic view of what is possible. Not all believe that de-growth can be done in an egalitarian way that retains the freedoms we now enjoy. Others have pointed out that de-growth movements put all forms of growth together in a way that does not distinguish that whilst some forms of growth are socially and environmentally harmful, others might not be. However, even critics of de-growth are often in agreement that measure of growth such as GDP are not sustainable and that consumption must be reduced in some sense. The question is then, can we agree on where to draw the line between harmful and helpful consumption? Once again this appears to be a question of balance, but beyond obvious examples of harmful consumption, such as weapons of mass destruction or addictive drugs, it is difficult to be objective in this regard. These are moral judgements that must be agreed across cultures and beliefs. The best guides we have are to look to those already living sustainably in terms of the environment and cross-check with those living the kind of lives that are also healthy, enriched and meaningful.

Fortunately, we need not fly to elsewhere to understand where other cultures are succeeding where we are not. Many studies exist that correlate human well-being with low consumption at a national level. At a more individual and personal level, we might simply need to be open to more honest conversations that challenge our prejudices and susceptibility to exploitation. For this to be a useful exercise we need to recognise the assumptions that many of us have been making, for no fault of our own, that the Western industrialised model is the path to happiness and fulfilment. For those of us who have cultural roots in less developed countries, this might actually be harder to do, despite the access to drastically different cultures we have. The colonial and post-colonial eras saw new markets opened up by encouraging a sense of inferiority that fuels aspiration towards Western urban development models, to Western design and aesthetics and to Western notions of beauty and superior culture, often in the services of economies elsewhere. This trend, though not complete, is still present in the lives of many next-generation immigrants coming from the former colonies. Since economic growth requires everyone to be aspiring to be one step 'ahead' it is helpful then that these hierarchies exist at a global scale and are reinforced at the local level.

At a time when the Western consumption model is no longer possible for everyone, the answer is not to devise a newer, slightly greener version of this model. Instead, it is time to develop a new shared narrative and to reconceptualise our relationship to nature. We live in an era of rapid social, technological, political and climate change. The good news is, we are a very adaptable species and history has shown that we require only a decade or so to change our social structures completely. Since we may only have a few decades to do this before we are forced into crisis mode, this offers hope as does the fact that communities all over the world are instigating change. Though we may not always have the answers or the unity needed to ensure success, simply establishing inclusive communities offers the opportunity to connect individuals to global issues and for prejudices to be unpacked. The most urgently needed communities are those that accept but transcend geography, cultural barriers and national boundaries, to view our planetary boundaries as guidelines to a better way of life rather than limitations to be overcome in the quest for endless growth. By their very nature these communities would offer our best hope for flourishing in the era of climate change.

FLARING GAS

Moiz Bohra

I grew up as Qatar was growing increasingly wealthy, sprouting the first of the skyscrapers that now line the Doha shore, and making a name for itself on the international stage. The source of much of this new wealth was natural gas, whose production, liquefaction and export by ship to markets in east Asia and Europe, was taking place in the north-east of Qatar. My childhood was spent in company accommodation close to these production facilities, after my father started working for one of the gas companies. Driving back home on the unlit highway from an evening in one of Doha's new shopping malls, we would count the glittering points of orange light on the horizon – these were the flares at the gas production facilities, burning unused gas in flames that were several metres tall and visible tens of kilometres away. In my eyes, the permanence of these flares on the horizon represented Qatar's eternal progress and ever-growing wealth. There was no place in this world for questions about the impact of fossil fuels on the climate.

That all changed for me when I was preparing for an environment-themed quiz in high school. It was the first time I came across the concept of climate change and I was shocked. The predictions of melting glaciers, rising sea levels, fiercer storms, global heatwaves and prolonged droughts seemed like prophecies of a biblical apocalypse, juxtaposed with the soft orange hues of the calm night sky outside my window, the low clouds reflecting the light of the permanent flares on the horizon. Had I not taken part in that quiz, or developed an interest in learning about environmental conservation, I could have spent the rest of my schooling, and perhaps even university and beyond, oblivious to the impending catastrophe. This is unfortunately common among those living in the Gulf states.

In hindsight, it seems inevitable that I decided to study chemical engineering at university. Qatar was investing its wealth into higher education by opening local campuses of some of the world's best universities, thus training the next generation of doctors, scientists and engineers from amongst the local population. I can't deny, despite my awareness of climate change, that I aspired to the same goals as everyone else around me – a life-long career in oil and gas, an air-conditioned villa, a fuel-guzzling car and frequent holidays abroad. Were it not for a general slump in the energy industry when I was graduating, and a timely scholarship that allowed me to travel abroad for post-graduate studies, perhaps I would have had that life – my only response to the changing climate being to turn up the air conditioning a bit higher every summer. Instead, I received a doctoral scholarship from the Qatar Foundation, moved to the US, and then to the UK, to study how the Gulf states can face the challenges of climate change and embrace a more sustainable future.

As a researcher working on sustainable energy in the Gulf, I was elated to see earlier this year school students around the world protesting the inaction of adults in combating climate change and pushing this topic up the political and media agenda. As a researcher working on sustainable energy and preventing climate change in the Gulf, however, I was not surprised to see that there were only a couple of schools in the whole region participating in this global movement. In general, there is a lack of public awareness in the region about the impacts of climate change, and the role of fossil fuels, produced in enormous quantities by the Gulf states, in pushing the world to the brink of environmental breakdown. This is not accidental in countries where the government is the gatekeeper of information. Moreover, these states are increasingly using more of the fuels they produce – domestic consumption of electricity, water and transport fuels is growing across the region, and most of these resources are produced from fossil fuels (due to a lack of freshwater resources, most Gulf states rely on desalinating seawater for all domestic consumption, a process which uses large quantities of fossil fuels). This domestic consumption is driven by state subsidies on energy resources, as this is one of the ways in which rulers share the spoils of fossil fuel exports with their subjects, in return for political acquiescence. Over the last decade, global oil and gas prices tumbled from the heights that had allowed the Gulf states to amass their wealth, and along with the youth-led protests of the Arab Spring, forced rulers to confront the economic, political and social realities of their countries.

My research on a sustainable future for the Gulf states was raising questions along similar lines. Why was there an inertia when it came to investing in renewable energy infrastructure? Why couldn't energy subsidies be reduced? Why was there an insatiable need for foreign labour, especially for the building of infrastructure? Why was the private sector finding it difficult to be seen as a prospective employer by citizens? How could a growing youth population find domestic economic opportunities? Why was the West ambivalent about the increasing global influence of the Gulf states – deepening its economic and security links while papering over criticisms regarding human rights?

The answers to these political, economic and social questions lay in the historical development of the Gulf states, and the close links between state formation and the oil and gas industry. The Qatar Digital Library, the result of a collaboration between the Qatar Foundation and the British Library, is an archive of recently digitised material on the Persian Gulf region, from the India Office Records of the British colonial administration. Although the focus of my research was Qatar, the archive's records hold key lessons for the other states in the region that have followed a similar historic trajectory.

The land that is now Qatar has been an arid, rocky, desert for thousands of years – its limited freshwater resources, almost all of it in the form of groundwater, supporting a population of nomadic Bedouins who migrated into the peninsula to graze their herds and left when resources were depleted. These Bedouins traded with merchants from other parts of Arabia, Iran, and India, in seafront settlements that formed the seeds of the first cities along the Persian Gulf. The region was known for its pearls, collected from the shallow seas by fearless men, perhaps half of them slaves, who dived into the water for several minutes on a single breath, with nothing more than a nose-clip and a small net to gather pearl-bearing oysters.

The region was first occupied by the expanding Ottoman empire but came under British and American control after the first World War. The British established security and economic links with the leaders of influential tribes, and eventually created nation states, with borders and rulers, along the Persian Gulf. In the 1920s, the pearl trade in the Persian Gulf collapsed due to the commercialisation of cultured pearling in Japan – pearls could now be grown by farmers in controlled conditions, with greater efficiency, lower risks and a more uniform quality. The global depression in the 1930's exacerbated the economic hardships for the region's population. Were it not for the discovery of oil by American and British prospectors across the Persian Gulf during this period, the region may perhaps have languished in poverty in perpetuity.

Although the First World War's reliance on mechanised weaponry and transport highlighted the importance of controlling oil resources during wartime, the onset of the Second World War refocused imperial attention to the domestic front, and delayed the exploitation of oil resources by another decade. Immediately after the war, oil production ramped up across the region. This was seen as a priority by the colonial administrations. Generous concession agreements were signed between the rulers of the Gulf states and the British/Americans – including access to almost all land for exploration, unlimited use of water, and exemption from taxes, in return for military protection, fixed annual payments and small royalties per unit of oil exported. The oil economy formalised the relation between the ruler and his subjects – for the first time, citizenship and labour laws were passed, and visas were issued for foreign workers. Sailors and settlers from across the Indian Ocean world, who were once as at home in these lands as the Bedouins of the desert, were now split into categories of citizen and non-citizen. A person's tribal identity began to compete with a newly-formed national identity that was created with the distribution of state largesse through subsidies on resources, land grants and cash payments. The rulers considered it their Islamic obligation to distribute a portion of their economic wealth to the wider population. Oil companies were managed by Western engineers and staffed by Indian accountants and unskilled Arab labourers, who for the first time in their lives engaged in formalised wage labour. It was in response to protests by well-organised oil workers that the states banned political activity and trade unions. State apparatus such as a police force, and infrastructure such as roads, hospitals, schools and electricity plants were all created to serve the growing needs of the oil companies. National borders, which would have been irrelevant to Bedouin nomads only two decades previously, were now the topic of fierce negotiations between local rulers and the colonial administrators, due to the wealth hidden under the sands. These disputes extended to uninhabited islands in the Persian Gulf once offshore oil reserves were discovered. The first offshore rigs were constructed using the expertise of the last generation of pearl divers. Employment records from the archives indicate that the Qatari oil company also employed slaves until the 1950s, when slavery was officially abolished by the ruler.

I learned about the history of slavery in the Persian Gulf region at a museum dedicated to this subject – Bin Jelmood House, part of the Msheireb Museums in Doha. Muslim empires were engaged in the slave trade long before its institutional adoption by Western colonial powers. The Omani empire, which at its height extended along the East African coast, institutionalised the trade throughout the Indian Ocean world – people captured in East Africa were sold at slave markets – from the shores of Western India to the oasis settlements of the Persian Gulf. When the British began to exert influence in this region, they did not enforce their anti-slavery laws so that they could maintain cordial relationships with local rulers who benefited from the slave trade. Although the slave trade was officially abolished in the region during the 1950s, the strict conditions of employment imposed on foreigners since then, including the sponsorship system that restricted workers from switching employers or exiting the country without the approval of their sponsor, continue the system of labour exploitation to the present, where millions of men from South Asia and Africa continue to work long hours in the searing desert heat for low wages to construct the newest record-breaking buildings, roads and man-made islands across the region, while hundreds of thousands of women from Africa, South and East Asia work as domestic servants in rich households, and still do not receive the same legal protections as other workers.

Meanwhile, Western governments have placed issues of human rights lower in their priority lists, when compared to securing the supply of Middle Eastern oil and gas demanded by their populations. Guarantees of military support have been backed by the construction of Western military installations across the Persian Gulf, and close ties between domestic and Western armed forces. Soon after commercial exploitation of oil began, local rulers first demanded a renegotiation of the royalty amounts due to them, thus accumulating even more wealth. However, as documented in the archives, the British also played a key role in deciding how the states' surplus revenue would be invested – whether at home or abroad. Over the last few decades, international energy majors such as Exxon Mobil, Shell and Total, among others, have provided the planning, construction and operational expertise for oil and gas projects, in return for generous part-ownership arrangements.

It is apparent that the history of oil and gas production in the region has a visible effect on all aspects of Persian Gulf states: citizen – state relations, labour rights, international investments, domestic infrastructure, internal and external security, and access to resources. Given the global appetite for energy, including new markets such as India and China, there would be no reason for these states to deviate from their present trajectory of fossil-fuel exports, were it not for the biggest challenge of our times: climate change.

Compared to other parts of the world such as Western Europe, there have been fewer studies on the impacts of climate change in the Persian Gulf. However, existing research unanimously shows an impending crisis in the next few decades, if the world carries on emitting carbon dioxide at the present rate. Lower rainfall will increase dependence on already depleting groundwater aquifers, which makes them more prone to intrusion by seawater. Increased groundwater salinity makes it unsuitable for domestic agriculture, further challenging the food security of countries that are already heavily reliant on food imports. The Gulf states are heavily reliant on the thermal desalination of seawater, powered by fossil fuels, to meet their domestic water demands – this process involves the heating of seawater to produce salt-free water vapour, which is cooled and collected. Not all of the water is evaporated, however, leaving behind a saltier stream of water, known as brine, that is disposed back into the Persian Gulf. The shallow seas of the Persian Gulf are more prone to rising water temperatures, which, coupled with the increasing salinity of the sea due to brine discharge from the large desalination plants dotted along the coast, have catastrophically affected marine life. The collapse of marine ecosystems is having a direct impact on the yield of fisheries across the Gulf, which were already declining due to overfishing. The Persian Gulf is also susceptible to sea level rise, which would irreparably damage its low-lying cities. Climate models predict that by the end of this century, in a business-as-usual case, cities across the Persian Gulf will experience a deadly combination of high temperatures and humidity on average summer days, making human life impossible outdoors, while also affecting the operation of key technologies such as planes and trains. Migration of climate refugees from poorer countries such as Yemen and Iraq may add to domestic pressures caused by a rise in food prices and the unavailability of water.

While the Gulf states may be able to mitigate many of these climate impacts with their significant wealth – from buying up agricultural lands abroad to building heat-proof and flood-resistant cities, they also face the increasingly real prospect of permanently reduced prices for oil and gas across global markets. From Western Europe to China, countries are ramping up their investments in renewable energy such as solar and wind – producing more of their own energy and relying less on imports from the Gulf. The rise of 'fracking' in the United States has unlocked an enormous quantity of previously inaccessible oil and gas, making it a net exporter of oil and gas, and pushing down the prices of these commodities further. The electrification of the transport (electric cars) and heating (heat pumps) sectors is picking up pace, directly affecting the demand for transportation fuels and heating gas. The rise of storage technologies such as batteries might be the final nail in the coffin for fossil fuels – allowing renewable energy from the sun and wind to be stored and used when needed. Of course, some challenges remain – particularly in the aviation and shipping industries, which are expected to grow, and remain reliant on fossil fuels in the absence of greener technologies. Moreover, growing populations in West Africa and India, aspiring to amenities such as 24-hour electricity access, air conditioning and rapid transportation, will continue to drive demand for imported fossil fuels, but these countries will not hesitate to switch to greener technologies, powered by domestic renewable energy, as their costs continue to decline over the next few years.

The emergence of the US as a net energy exporter means that instead of relying on a secure supply of energy from the Middle East for its domestic consumption, it is now engaged in global competition with the Gulf states. For the first time in their history, states in the Persian Gulf must contend with a West that does not depend on them for its energy needs, and no longer needs to guarantee their internal and external security in return. We see this shift already taking place – some Gulf states are overstating their opposition to Iran, identified as a threat by the current American government, to remain under the security umbrella of the United States. In the longer term, however, all of these states will have to adapt to the changing geopolitical landscape of the Persian Gulf.

The multiple challenges posed by the domestic impacts of climate change and the long-term decarbonisation of global energy systems elicit the need for a systematic transformation in the Gulf states. My research aims to provide pathways for this change – by studying the suitability of various green technologies to meet domestic demands for electricity, water, transportation and air conditioning, by optimising the industrial export portfolio for a post-carbon global market, and by recommending policies that will enable this transition and secure the economic future of the region. While my research has focused on the optimal energy transition pathway for Qatar, many of these solutions are applicable across the Gulf states.

Perhaps the most obvious green technology for the Middle East is solar energy. Photovoltaic (PV) panels that convert sunlight into electricity have seen a significant cost decline, which is expected to continue. Per unit of electricity generated, they are already cheaper to build than gas-fired power plants, particularly for countries that do not have adequate domestic gas supplies. The United Arab Emirates has invested in the first large-scale solar PV projects in the region, and saw some of the cheapest electricity prices in the world when it called for tenders from companies to build these plants. One of the most important factors for these low prices was the presence of a centralised electricity distributor, usually the case across the Gulf, who can always guarantee the purchase of solar power, as opposed to other parts of the world where the grid may not be able to accommodate solar power whenever it is generated. Moreover, solar PV produces the most electricity at midday, when there is also peak demand for electricity for air conditioning, thus reducing the additional cost of ramping up the output from back-up gas-fired power plants to meet the daytime peak. The falling costs of grid-scale battery storage will complement the introduction of solar energy, as excess electricity produced during the day can be stored in the batteries and used to meet evening demand for electricity. Apart from grid-scale renewable technologies, homeowners may be incentivised to install solar panels for electricity, solar water heaters and batteries to store excess electricity. This will only work if the government pays for these systems, in exchange for the reduction of electricity subsidies, to incentivise renewable generation in homes. A regional high-capacity electricity grid can also even out the variation in renewable energy production, allowing countries to buy and sell electricity. Rather than individual air conditioners in homes and offices, new parts of cities must have district cooling systems, which are centralised units that pump cold water through pipes into neighbouring buildings and satisfy their air conditioning needs. Due to their centralised nature, district cooling systems are more efficient than individual air conditioners and are already being adopted in several new urban developments across the Gulf. In addition to green technologies, governments must also mandate compulsory energy efficiency measures, which have seen limited voluntary adoption due to heavily subsidised domestic energy prices. These include better insulation for buildings and higher appliance efficiency standards. Subsidies on energy must be slowly phased out, as these are an inefficient way of wealth redistribution and have shown to benefit rich households over poor ones – direct payments to poor households are a more efficient form of redistribution.

Transportation infrastructure also needs an overhaul. Cities in the Persian Gulf were often modelled on American cities such as Houston, with miles of roads cutting through urban sprawl and limited options for public transport. Since cars will continue to be an important mode of transport in the future, electric cars, increasingly considered as luxury products and thus seen as a status symbol, can be promoted by expanding charging infrastructure. More importantly, public transit systems need to be affordable and accessible to most of the population – this means that stations should be within walking distance, or connected by a reliable bus network, from residential and commercial areas, and tickets should be cheap enough to provide an affordable alternative to driving a car or using a taxi. Systems such as the Dubai Metro that are well-used by residents must be expanded, while similar transit projects introduced in other big cities across the region, such as the under-construction Doha Metro. The reduction of transport fuel subsidies (that has been occurring across the region in the last few years) will be easier to implement if the population has access to an alternative means of transport such as a convenient and affordable metro system. Another opportunity for decarbonising transport lies in the freight sector – large trucks running on diesel carry almost all goods, clogging up roads and increasing air pollution. New technologies such as hydrogen-powered freight trucks, close to commercial implementation, may provide some benefits, while the construction of a regional high-speed rail network, previously proposed but hampered by political disagreements, can facilitate the transport of goods and people across the region, thus also reducing the reliance on air travel. The decarbonisation of the transport sector will also have a positive impact in reducing urban air pollution, which currently affects the health of millions of people across the Persian Gulf, and disproportionately affects younger and older people.

The lack of freshwater resources in the region means that the Gulf states must rely on desalination plants to meet domestic water needs, but the impact on the natural environment can be mitigated by mandatory water conservation measures in the residential, commercial, industrial and agricultural sectors, reducing losses in water transmission, and switching to reverse osmosis desalination – a commercially-viable technology that relies on electricity rather than fossil fuels to separate salt from seawater, potentially drawing its power from solar energy. Urban and rural areas also waste a lot of desalinated water for the irrigation of green spaces and farms using inefficient techniques. These losses can be reduced through better irrigation systems (such as drip irrigation, where water is directly supplied into the root zone of individual plants, reducing evaporation losses) and the use of treated sewage effluent for irrigation, as is increasingly being implemented across the region.

In the current environment of low oil and gas prices, countries reliant on hydrocarbon exports may no longer be able to justify the expansion of the energy industry. In Qatar, the production of low-carbon hydrogen (from the chemical conversion of its abundant natural gas resources, which also produces carbon dioxide that will need to be stored underground) might drive the future of the energy industry, particularly as demand for the cleaner fuel is expected to grow in countries like Japan, which has been a key customer for Qatari gas. Countries are also keen on expanding the production of petrochemicals such as polyethylene plastics – however, global efforts to reduce plastic consumption might challenge such projects. The use of renewables to meet demands for heating, cooling and electricity can also reduce the carbon footprint of industries. As in previous decades, governments will also have to negotiate better profit-sharing arrangements with international partners, or perhaps use their accumulated expertise to plan, build and operate their own industries, independent of foreign companies, thus keeping a bigger share of the profit. More radically, Gulf States could focus their efforts to reduce their economic dependence on hydrocarbons by setting goals to reduce production over time, perhaps going to a target of zero fossil fuel exports by 2050.

These investments in urban, rural and industrial infrastructure can create substantial employment opportunities at all skill levels for the local workforce. If these projects are government-led, they can also provide the job security and other benefits traditionally associated with public-sector employment. Projects such as a long-distance rail network and a regional electricity grid also open avenues for increased collaboration between countries – which has historically been lacking. This can lead to a deeper economic integration of the Gulf and reduce political friction. The large-scale transition of urban infrastructure may also spell the end of flashy projects such as new urban districts on artificial islands – these projects have been prone to cost overruns, delays and excessive energy consumption, not to mention labour exploitation and under-utilisation once constructed. Instead, infrastructure investments need to focus on adapting cities to meet climate threats such as high temperatures and rising sea levels. These efforts must go together with educating citizens, especially the next generation, on the impacts of climate change and the scale of the transformation needed to prevent its worst effects.

The decoupling of the economies of the Persian Gulf from the energy industry will necessitate greater fiscal accountability. Citizens must be involved in the decision-making around how revenues from fossil fuels must be invested, so that they are onboard with the decarbonisation of their country's economy, knowing that the wealth that is invested today will provide their children with continued prosperity in a post-carbon world. The sovereign wealth fund of Norway, which is responsible for investing the country's oil revenue for the best interests of its citizens, is an excellent example of an institution making transparent decisions and sensible long-term investments. The proceeds from such investments can be used to provide a universal basic income for all citizens in a post-carbon economy, who would then be free to pursue their creative and entrepreneurial dreams, without having to rely on unfulfilling public-sector employment.

The transition to a post-carbon future gives us the opportunity to create a more equitable and just world. Qatar has implemented projects, over the last few years, to reduce the flaring of gas in its industrial facilities. Those glittering points of orange light on the horizon, an enduring image in my memories, do not have to be a permanent part of Qatar's future.

'LITTLE LONDON'

Muhammad Akbar Notezai

It's the fourth day of Ramadan in the month of May. It's 10.30 in the morning and the temperature in Quetta, the provincial capital of Balochistan, is soaring at 27 °C. People are fasting and there's little traffic on the roads. Twenty-seven degrees Celsius for a city in the southern hemisphere shouldn't be unusual, except that we're talking about Quetta. Known by older generations as 'Little London' Quetta is famed for its greenery, abundant valley freshwater and fancy whitewashed British colonial-era architecture. Its late 1800s-built train station wouldn't look out of place in an English town.

Today, Quetta and its environs are far from this possibly idyllic past. Quetta's population has soared to nearly 3 million, average temperatures are on an upward climb; crops are being harvested at weird and unpredictable times. A pattern of stop-start rainfall is creating famine-like conditions in one year, and flood-like conditions the next.

A decade ago, temperatures averaged 28°C. That figure has jumped to 32°C and is rising still to potentially unimaginable highs. If we just take the month of May, most of Balochistan's drier districts have received 4 or 5mm of rainfall every year since 2009. But last year, the entire month saw slightly more than 1mm in the province's west. Compare that with March, which saw a deluge of 100mm, instead of the 20 or 25mm that is normal for that time of year, and this picture of unpredictability starts to become clearer.

Quetta, its people, and its neighbours in Balochistan's more arid regions are grappling with the effects of an environmental crisis, the likes of which have probably not been seen in most lifetimes. To better understand just how serious the situation is, and what the causes might be, I am heading 400 kilometres further west of Quetta to a region of Balochistan called Chagai.

Type the word 'Chagai' into a search engine, and the results also include the words 'atom bomb'. That's because this hot, dry and seemingly desolate area, close to Pakistan's border with Iran and Afghanistan, was the site of a series of underground nuclear tests in 1998.

Two decades ago, almost all of Pakistan wildly celebrated as the nation joined the ranks of the world's nuclear powers. Thanks to YouTube, fans of the bomb can endlessly replay the magic moment when the Earth shook from the force of nuclear-induced seismic waves. But those TV images of jubilant scientists surrounded by miles of desert always hid an uncomfortable truth: Chagai is not an empty wasteland of nothingness. It has always been home to people, to crops, to domesticated animals and to thousands of species of wildlife. Today, close to a quarter of a million people live here, and most are very, very poor. For generations, successive governments have neglected their needs, and now it seems they're losing access to their last remaining lifeline: water.

My first stop is Dalbandin, Chagai's administrative headquarters. The temperature has already shot up 11 degrees to 38°C though there are clouds overhead and they briefly open up for a welcome light drizzle. I'm meeting Khalil, a young Dalbandini boy with hair parted down the middle. He's wearing Baloch-style shalwar-kameez over a sports t-shirt. I'm going to join Khalil on his monthly mission when he delivers a ration of supplies to his uncle Abdur Rehman. Now in his seventies Abdur Rehman lives alone, twenty-seven kilometres northwest of Dalbandin just outside a nest of six villages that go by the name of Kani.

I have hired a car for this part of the journey and forty minutes later, we find Uncle Abdur Rehman. He's sat outside a tattered old hut where he's mending a boot. Some of his clothes appear indistinguishable from the colour of the surrounding land, torn by constant wear and beaten up by the unrelenting climate. Although it is the holy month of fasting, Abdur Rehman tells us that he is often perpetually fasting because there are days and nights when there is nothing to eat. Today is one of those days, which is why he is relieved to see his nephew.

Abdur Rehman occupies two huts: one for himself and one for his dozen goats. 'Why did you come in this weather?' he quizzes us – the rain has now given way to winds. 'You could have come in the evening or in the morning, when weather gets normal?' he says in a kindly way.

It turns out that Abdur Rehman hasn't always lived in such visible poverty. He had six siblings, but most have since died. He owned land too, where he would grow water melons and various varieties of grain. But the land is now too dry to cultivate; and the lack of pasture has also meant that his once-impressive herd of 400 goats and sheep has dwindled to a dozen. 'One by one, they died from drought. I could see them dying', he recalls. 'People from Kani and from Dalbandin would come here to take milk and butter from me,' he says. Now, the lack of even the most basic amenities has made him vulnerable to theft. Abdur Rehman complains that local youngsters come to his hut looking to steal sugar and other items from his monthly ration.

The water crisis has seen Kani's six villages gradually depopulating and numbers are now down to the low hundreds. Many families have moved out to Dalbandin, from where they will send supplies including rations of food and water to those like Abdur Rehman who are staying put. 'He lives for these goats,' Khalil says. 'Sometimes if there's storm, or rain, or if the weather is unpleasant and windy, he brings the goats to his hut. The whole night he will sleep alongside his goats.'

We exchange pleasantries, then Abdur Rehman briefly disappears. He returns laden with sticks of dried timber. A fire is lit and he extracts a teapot from a nearby sack: there's barely enough water for one man to drink, but Abdur Rehman thinks nothing of sharing what little he has—along with memories of better times, with his two visitors. 'I'm sorry I am serving you black tea,' he says apologetically. 'The goats do not produce enough milk.'

Balochistan is one of Pakistan's five provinces. It's the largest by land-area, the smallest by population. It is the richest in terms of reserves of coal, minerals and it supplies the entire country with natural gas. And yet its people are among the nation's poorest. That is partly because of history: after independence from British India in 1947, a few of the province's feudal rulers did not immediately embrace their new Islamic Republic and for several years operated as a semi-autonomous region called the Balochistan States Union, or BSU. But Balochistan is also poor because few of the benefits from its natural resources have flowed back to its people. Astonishingly Islamabad's policymakers had no qualms in sucking out the gas from this gas-rich province without giving Balochistan's people access to the fuel for their cooking and heating. That combination of poverty and official neglect has made the population more restive than before.

That said, it would be unfair to blame government officials for all of Abdur Rehman's predicament and that of tens of thousands of families like his. There are government officers who live in and around Balochistan's drought-hit districts. They live and work in the same communities as everyone else and they are every bit aware of what is happening – though largely powerless to do anything other than monitor and measure. They include Abdul Nasir Bangulzai, the Officer-In-Charge of Chagai's Meteorological Department. Bangulzai confirms that the total quantity of Chagai's annual rainfall has been dropping, and he says that there's no doubt that the lack of rainfall is contributing to surface temperature increases. He is also convinced that climate change is the cause. 'If it continues to remain so, it will have an adverse effect on animals, plants, humans in the future,' he predicts.

Balochistan's human victims include children and many now face malnourishment, explains Chagai's District Health Officer Dr Imdad Baloch. In a recent survey, Dr Baloch's office examined 6,000 children. Close to half he found were malnourished, while a little over a thousand were extremely malnourished—in other words, they were close to losing their lives. 'These children are mostly aged between 5 and 10,' he explained. It's not much better for their mothers. Of 388 mothers examined in the same survey, 168 were found to be severely malnourished.

Then there's the impact on agriculture, which is how most of the population earns a living and feeds itself. Four out of five wells and springs have dried up, and 11,000 acres of what was once arable land is now too dry to grow crops, says the Agriculture Department's Khalil Ahmad. This he says is leading to the spread of diseases among plants and especially animals as they lack access to water and animal feed. The Livestock and Dairy Department's Deputy Director Dr Saeed Baloch estimates that more than 200,000 cattle are in poor health. 'Some goats are so weak that they cannot stand,' he says. Most resort to eating grass, which is not a long-term diet, and that, too, contributes to early death.

Poverty, illiteracy, infant mortality and childhood malnutrition have always been unacceptably high in Balochistan. So is it fair to say that human-induced climate change is the cause of renewed suffering? That's the question that I set out to ask a group of Quetta-based writers and academics.

Science writer Safiullah Shahwani certainly thinks so. 'There's a greenhouse effect,' he tells me. 'There are emissions of carbon dioxide; there are industrial activities; these are global phenomenon. Most places in the world are already affected. One of those places is Balochistan.' He talks movingly of the 2010 floods. 'People died; people migrated from their ancestral towns in thousands. And that happened due to climate change.'

But Saadeqa Khan, a young novelist and science journalist, isn't so sure. Saadeqa edits a science magazine called *Scientia* and her novel *Dawaam* is about natural disasters. She tells me that Balochistan has always been drought-prone, and that drought cycles occur every five to eight years. That said, she is sure that deforestation is also creating the conditions for broader environmental change. Abdur Rehman is not alone in needing to cut down trees to light his fire. When tens of thousands need to do the same, that represents a serious environmental problem.

It's nearly time to wind up my short visit, but before I do, I need to make one final call. I am on my way to the University of Balochistan, situated in the south of Quetta, and to the office of Professor Zahoor Ahmad Bazai, one of Pakistan's leading environmental scientists. Professor Bazai studies the effects of pollution on food. He once gave a talk in Germany explaining that Quetta's vegetables are grown entirely from contaminated water. At the end of the lecture, another professor stood up and asked to know how he was still alive.

Why is Quetta becoming hotter and drier, I ask Professor Bazai? To what extent is climate change the cause, I want to know. 'Undoubtedly, climate change has become a reality now across the world,' he says. 'It's not just me who says so, but [even more] renowned scientists,' he asserts in a self-deprecating way. 'The percentage of CO_2 emissions is on the rise, which is disastrous.'

Ultimately, Professor Bazai says, the solution to climate change involves resolving the fight between business and the environment. Business-people – among who he includes US President Donald Trump – are resisting climate change mitigation because large industrial corporations and smaller ones stand to lose money by changing how they extract and use energy. Some are coming round to the idea that action needs to be taken, but the reality is that it may be too late before the majority mend their ways, or change their minds

It will almost certainly be too late for Uncle Abdur Rehman. It's time to say goodbye and he comes out of his hut to bid farewell. He has never heard of climate change, nor should I or anyone else expect him to describe the science of global warming. But Khalil's kindly old uncle is more aware than any scientist that his environment is changing, and that the effects for him and for his family have been catastrophic. He has lost the use of his land; he has lost most of his livelihood; and now he is at severe risk of losing access to water.

'One day the people of Kani will find me dead,' he says. 'But I will be surrounded by my goats.'

BABEL BRICKS

C Scott Jordan

Here it is. My shot. A perfectly thrown pitch, begging me to knock it out of the ballpark. Climate change. Truly this is the fundamental struggle of our age, whether or not we are willing to accept it. Yet here I sit, before my laptop. My brain is racked. I am defeated. What more can I say of global warming that the healthily grown tome of postnormal times writing hasn't already touched upon. What cleaver path remains uncharted? I could speak more to what we are all witnessing. We are slowly being cooked alive in our homes, if they are not being cast off in the rising sea levels. Season's change is, grossly put, irregular leading to a general daftness in the poor fauna, no doubt including many humans. The increase ferociousness and creativity in mother nature's path of destruction. I could shout solidarity with pride as I watch the innovative robots that pull plastics out of the ocean, the creation of the youngest generation, still yet retaining a sense of curiosity and imagination that will be robbed of them in the coming years. Included in this generation are the students we see flocking from their schools around the world to protest the global lack of action concerning global warming. Though it is unclear how much of this is populism turned to the forces of good and how much is composed of those looking for a day off from the mind-numbing indoctrination and innocence robbing machine of contemporary education. Yet, as my face goes the azure hue, I wake to the sounds of construction that sing me to sleep. I listen as horns and muffler-less motors roar along the congested streets. I watch as bag after bag is filled and mild attempts at recycling are confounded when all rubbish is tossed in one collection. In our hearts the planet's concern may lie, yet we consume at a rate that would even have the Coneheads begging for moderation. So, in the end, as the author George R. R. Martin loves to recite in his *Song of Fire and Ice* book series, 'words are wind…'. And if they could only be collected by wind turbines, we would have a new standard of renewable energy.

Brought for your convenience by postnormal times, here we are in a great simultaneity which reorients our words back at each other. As I sit down to write this, I watch as two good friends, one of Pakistani and the other of Indian descendance, argue online over the revived Kashmir dispute. Reason abandoned, calls for blood and fire underlie very cliché, yet clinically dangerous, examples of uncertain revenge and nationalism. In the UK, the two major parties, the ebb and flow of British democracy are literally dissolving before our eyes as the second-hand ticks towards midnight on one of the most catastrophic global moves in recorded history. In the US, President Trump claims the construction of a wall on the southern border a matter of national emergency, over the rapid descent of the Midwest below the flood's path, and that which he is willing to stake his presidency on. Let's not forget Trump's choice of William Happer, an adamant global warming denier, as his chief advisor on 'climate change'. Fascism is back, and almost as ubiquitous as Global Warming. And all I need to do is utter the words 'Middle East', and perhaps you see where I'm going.

So, what am I to do when the world gets so nauseatingly loud? What I always do in times of trouble. Find a nearby cinema and with the purchase of a ticket, muffle the noise a bit, which, in most cases, is enough.

Less than distant memory gives us a standard for films concerning climate change. They even follow a fairly universal structure making them easily consumed by audiences without much need for thought beyond the simple somatic process of lifting popcorn to mouth hole. These are the flicks given to us by such American 'filmmakers' as Michael Bay and Roland Emmerich. Essentially, we are given a shabby every-man protagonist who has a world of his own problems. There is often a half-cocked attempt at a strong female supporting character that has a rich knowledge in science, great for pesky exposition, or fervently independent. Unfortunately, this progress is stifled by that character either having a job in the sexual entertainment field or tragically fated to the damsel in distress trope. Often, there is a curious child who has just the fresh perspective to save the day in the end. Then by the inciting incident the world is plunged into chaos by a natural disaster that we should have all seen coming, but here we are, caught with our pants down. Our every-man stumbles his way into saving the day and even the cute puppy as well. The second act has space for freedom to insert lovely bits of nationalistic propaganda, heart-warming comic relief, ham-fisted political rhetoric, and maybe a budding romance. By the third act, the country, and generally the rest of the world, is left in utter ruin. But then, usually a broken politician, or father reborn in tragedy a better individual, gives a resounding speech of how we stood together as humanity and defeated the threat before us, now with lessons learned, together we build a new world, blah blah blah. Credits. A smorgasbord of special effects accompanies a lesson from your choice of parenthood, maturity, individualism and courage. Munch munch munch goes the popcorn. After leaving the cinema, we go about our classically destructive lifestyles. If you wish for a bit of homework, apply this formula to such films as *Deep Impact* (1998), *The Day After Tomorrow* (2004), *2012* (2009), *San Andreas* (2015), *Geostorm* (2017), to spot a few from within the litany.

This changed with the often under appreciated recovery from the Batman trilogy, Christopher Nolan's *Intersellar* (2014). In postnormal film, the idea of a future where humans continue to live on Earth has largely become unthought. It would appear at a glance that global cinema is also carrying forward this hopeless view of Earth's future. The South Korea-Czech *Snowpiercer* (2013) and China's *Last Sunrise* (2019) are tales of humanity's continued struggle through the next ice age, attempting to hold onto the last vestiges of the familiar. Moving beyond earth as *Interstellar* advised, Sweden's *Aniara* (2019) looks to Mars, adding to the wonderful genre of sci-fi horror when the titular craft falls off course en route to the red planet. China's *The Wandering Earth* (2019) combines all of this and maintains the formula of the typical climate change popcorn flicks. In this film, faced with the threat of an expanding sun, the objective is to move the Earth to a new solar system. To do this, the Earth's rotation is ceased, inducing an ice age and forcing the population to burrow below the surface. Rocket engines are then constructed, largely in the Asian hemisphere, to propel the Earth. The trickiest point in their navigation is avoiding a collision with Jupiter. Our every-man, our failed attempt at a feminist co-lead, and our goofy comic relief must prevent this catastrophe. Meanwhile, the father of our every-man is leading a multiple decade long mission, a craft with hibernating humans, that will seek a new home if

Project Wandering Earth is to befall a tragic fate. It is interesting to note the general themes pervading postnormal climate change films. In *Interstellar* and *The Wandering Earth* we have fathers attempting to make the situation better for their children, a sentimental commentary on the present's failures to the future. While a global government appears to be a staple of these visions, nationalism and patriotism remains intact, if not magnified. It appears the human element is all that remains, even when mother Earth herself is to be cast asunder.

It should not be underscored, that there is another class of climate change-preoccupied film. The message is far more subtle than explosions and cataclysm. *Beast of the Southern Wild* (2012) is a wonderful fantasy playing out in an innocent young girl's head, but set to the backdrop of the first major victims of climate change, who will inevitably be the poor and disenfranchised. They are to become the first wave of climate change refugees. Bollywood has a long repertoire in this tradition with *Mother India* (1957), *Bhopal Express* (1999), and *Jal* (2013) where the effects of climate change, often in a not so subtle fashion, act as grand antagonists. In following the tradition of *Bladerunner* (1982) and *The Haunting of Hill House* (2018), where the setting itself is lead character, climate change will continue to build up its IMDb page in the coming years.

A rich gallery of images has been produced. But spectators should be advised, that following its viewing, despair, depression, and hopelessness are expected side effects for those who think beyond the spectacle. We can see judgement day, but can we change? Can we prevent such destruction? Can we transcend global warming and climate change? Even, the supposed ultimate escape of the contemporary superhero film is stained with the hopelessness of climate change. And just like our brave heroes at the end of *Avengers: Infinity War* (2018), we haven't got a clue. With its follow up being subtitled, 'Endgame', matters do not look good.

Gorging myself in such films has left me with a bit of a belly ache. I needed something different. So, I rolled the dice and sought out a small independent cinema.

In the darker than normal theatre I am not accompanied by my usual accoutrements, that of a nice bag of buttery-salty popcorn and a refreshing fizzy drink. No, this moviegoer outing is serious business. Oscar season is at hand and a number of foreign language films have made the cut. For a film that I need to read as much as watch, my greatest co-pilot is a large black coffee, sans milk, preferably prepared at temperatures that would make Guy Montag sweat. The chair is comfortable enough, but in complete blackness a red flag is thrown at me. There are no adverts for other films or anything accompanying the main event. No time to adjust, crack open your munchies or clear your throat. I am alone in the room, not necessarily an uncommon thing in this age of streaming, for after all this is no superhero blockbuster. So, I alone am thrust into the film with no warning. But I am prepared as I sip my coffee.

Stark black and white often oozes pretention, yet here it is justified for the film takes place in the late 1940s in post-war Poland. The composition of the opening shots makes the photographer within me carve a smile upon my face. A mild discomfort grows into anxiety as the opening shots, of people singing, are not accompanied with subtitles. Fair enough, stay calm, the canvas is simply being set. Your ignorance will be acquiesced momentarily. Yet as the various groups and individuals singing and playing instruments, two characters are introduced recording the music and dialogue takes place between them. But their conversation seems unimportant and they are drowned out by the music they are recording, so perhaps, I the audience am not to know what they say. I doubt the film's integrity hinges on this. It takes about fifteen minutes to realise the version of this film has no subtitles. At first, myself having a fragment of the collage that is my heritage being from Poland, feel a hint of shame that I do not even know a word of the language. This anxiety mixed with my respectful desire not to disrupt a presentation over my own personal lacking of understanding, forces me to buckle in for whatever is in store for this filmgoing adventure. After all, I lavished in the beauty of the Ukrainian film *The Tribe* (2014), entirely performed in sign language, where the filmmaker expressly left out written subtitles.

Luckily the many title cards inform me that we are moving between Warsaw, East Berlin, Paris, and Yugoslavia throughout the 1950s. A great deal of credit must be given also to the cinematographer for filling the frame with symbols that keep me completely aware of the theme of the dialogue that I cannot even begin to understand. For instance, when a bust of Vladimir Lenin stands on a table in the lower corner of the frame I begin to pick up on the Polish equivalent of 'proletariat', 'bourgeoisie' and other communist buzz words. Likewise, when I see the progression of dance performances move from rural 'old world' Poland in their costume design to suddenly having standardised uniforms and the giant head of Stalin behind them, I see what the point of these performances. The continued people's revolution! A bottle of vodka ought to have a starring credit, but this symbol nicely sets the mood of the Soviet bloc at its peak. A tastefully composed sex scene, reveals to me that a love plot is afoot. The repeated setting of a bombed-out eastern style church and the image of a biblical figure, most likely Jesus, veiled but for the eyes in soot, provokes deeper interpretations of love, humanity, and justice amidst the dividing backdrop of politics, sectarianism, and the Cold War.

Upon a brief discussion with the proprietors of the cinema house upon the film's conclusion, I learned that some mix up in the projection room contributed to my experience. Despite the profuse apologies by the proprietors I assure them the lack of subtitles was not a problem at all. My ability to understand *Cold War* (2018) without the dialogue speaks to the value of the piece and the magic of motion pictures. In fact, upon further research into the film, I believe that had I watched this film with the proper subtitles, I would have appreciated it less, as I am not a fan of simple stories of lovers caught between the divides of human politics.

Cold War launched me on a furious journey to watch all of the top foreign language films of 2018 as so deemed by the Academy of Motion Picture Arts and Sciences. First came Alfonso Cuarón's *Roma*. The film follows Cleo, a maid to a higher class family in 1970s Mexico, as we see her humanity stripped in the background of Mexico itself descending into political chaos. Japan's *Shoplifters* tells of a band-turned-family of thieves exploiting the struggles of fringe society in contemporary Japan when the 'son' is arrested by police. Lebanon's *Capernaum* is the story of a boy suing his parents, following his own tumble into depravity, for the horrible world they have left him to inherit. Germany's *Never Look Away* tells the story of an artist coming of age in Soviet occupied East Germany obtaining his freedom in the Western sanctuary of Dusseldorf, yet retaining the stains of Germany's lost pot-war generation in his mind and in his work. Even Sweden's *Border*, which got an honourable mention, gives voice to mystical creatures as they comment on and are confronted with the evil nature of humans in the contemporary world.

The themes of these films ring clear to the contemporary challenge we are faced with despite their linguistic and temporal differences. The underlined irony of the upper classes being more depraved and uncivilised than the dirty and scraping to survive lower class individuals and the stains of past failures infecting the present minds of the youth speak to something that climate change films ought to take note and learn from. For perhaps our creativity can be focused into taking us beyond the horrifying end and working against such visions, empowering the present to take control of what futures may lie before us.

As I dived deep into 2018's non-English films, I watched old climate change documentaries and was struck that the human element, presented in focus and high resolution in the contenders for the Best Foreign Language Film Oscar, was lacking. Naomi Klein presents in her book turned documentary, *This Changes Everything,* that the issue with climate change is that we need to alter the narratives of global capitalism and consumption in order to save the planet. Yet she offers no alternative. She portrays the people suffering on the ground, yet her message remains hopeless for the powers at be, profiting off the destruction of the Earth, are too powerful. The documentary *The Devil We Know*, is an expose that attempts to be a bit more hopeful by taking on the DuPont corporation who had been knowingly dumping the harmful C8 biproduct formed by the reaction that produces Teflon into the water supply. The tale told is of locals banding together to legally take on the devilish corporate overlords in the face of popular dissent and fiscal pressures. A brilliant sequence is dedicated to the local farmer Wilbur Tennant, living just downstream of the DuPont plant where C8 dumping began in the 1960s. Using low quality video recorder accompanied with Tennant's unscientific commentary on how dead fish are washing up on shore and his cows are developing a black build up around their teeth before dropping dead or giving birth to deformed offspring. Following Tennant and his wife's passing shortly thereafter, due to cancer, likely a development of drinking from the contaminated water, his case was quickly settled and swept under the rug. Had Tennant only lived in the current age of social media and his message allowed to travel at the speed of tweet, Dupont would likely have suffered a similar pain to the swatting of a 1980s sitcom father's heads with a Teflon-coated pan wielded by the social justice warriors of today. Sadly, this story ends with the struggle of our locals continuing to this very day against Dupont. Only a microcosm of the greater issue was left between the credits. The quest for solutions seems at this point futile.

During Klein's documentary there is a sequence where the filmmakers attend the 2011 Heartland Institute Conference. Despite the legitimate sounding name of this event, at it core, it is a place for climate change deniers to assemble and denounce the false propaganda of the tree huggers! They are another of those pro free market think tanks, but have found a peculiar talent for denying the scientific evidence for climate change. They yell socialism when they see green proposals. In this documentary, I was taken by two particular comments made by one of their representatives. The first was, if you want more trees, buy more wood. The buying of wood will trigger the market to demand more trees. Likewise, if you want more elephants, buy more ivory. The inclusion of this person's quote was intended to illuminate the absurdity of the Heartland Institute. While this was accomplished masterfully, there is something deeper at play here. We can sit here and say, look at these absurd people until we go blue in the face. But these people and their bankrollers actually believe this is logical thought. Their worldview is neoliberal economics. Free market or bust. They are followers of a new god, the market. This is similar to the people who believe recklessly in the Earth's power. They will say, what ever we do to the Earth, it is strong enough to overcome. They neglect to emphasise that from this logic follows, that the Earth will defend itself, even if that means ending the unfortunately brief encyclopaedia entry that is the life and times of *Homo sapiens*. For them, if you want your green agenda, only the almighty market can bring that about. Therefore, the only way to change those of this ideology is to combat their logic with their own language. Climate change illuminating film, thus ought to be revealing that the market has no clothes, in the framing and terminology of neoliberal economics.

Speak another tongue, and the deniers will rally behind their misunderstanding and disregard you for your foreignness. So how do you speak to someone who is attempting to kill you? Old school political types will tell you this is an exercise in futility. We have seen how far along the road of progress it has gotten them and us. Do we continue fighting the great fight *ad infinitum*? After all, the *Star Wars* cinematic universe seems to be the same resistance against the evil ones in perpetuity, and have we not grown tired of that conflict? Forgive my brutal language, but climate change is the war of our times. The battles we fight today will define the millennial generation and while ideologies come and go, there is only one planet Earth. And our imaginations can only hope there is another rock like it out there in the cosmos. A united human front is the only way to save this planet and that is non-negotiable. A united effort requires, if not a united language, if not a united understanding of climate change, and if not even that, at the very least a clever bit of translating that allows us to transcend our logic and faith structures. Unfortunately, this is a zero-sum game. Unlike wars of the past, where only a few major players determine the flow, and some can even remain neutral, if one element does not cooperate, the entire complex system is thrust into chaos. From a global vantage point each of us is a member of an archipelago who's tether wears thinner and thinner by the tweet.

History can give us some directives to correct our tectonic attempts at hermitage. The flavours of ethnicity and nationalism that have occupied Palestine and the Balkans, two among many similar international situations, can teach us about the emotional and illogical drives behind enmity. Being a child of divorce has taught me a great deal about the extent to which two individuals can feel hate for one another despite being tied to something primal to their existential being. As the trend of divorce appears to be growing perhaps many more future children will join my classification. At this point though, we must ask, do we really want to wager all in on a global generation of children of divorce to see if that's how we can all learn to get along?

Of course, the situation goes beyond enmity and history. Perhaps I am being a bit dramatic when I label climate change deniers as those who wish to kill me. Their intention isn't to kill me, its just that their actions circumstantially result in my death. Once upon a time, I found myself a part of a reading group while I was in university. The reading group was launched by a respectable enough individual who was also a member of a centre paid for by the Koch brothers, two of America's richest supporters of right-wing causes. The reading group was established, partially, to quell the controversy of our university having such a research centre. My inclusion in this group was, I have come to believe, to further press this point. My job within this reading group seemed to be to put undergraduate business degree seeking students firmly in their place, give the eleventh man position on neoliberalism, and quote Karl Marx once or twice a week. Now I was quickly labelled the Marxist, which added a pesky, demeaning footnote to every comment I made. Though I was quick to note that I was not a Marxist per say, just an individual who had read *Capital Vol. 1* from cover to cover and analysed many semesters worth of his essays and manuscripts. I would kindly ask that they correct my label to being the 'Marxist bastard' of the group. So, there I was reading Thomas Picketty's *Capital in the Twenty-First Century*, and a scorpion's nest of contemporary pseudo-economics and classical economists of the Austrian School trying to find places to correct for inequality. As if neoliberal economics were just a code of DNA needing a few deletions and insertions. Meanwhile I became the court jester, who attempted to reveal the fundamental flaws of neoliberal thinking via mockery and quote and verse reprisals of Smith, Marx, and Engels. At the end of the day I can only hope my fellow readers at least read Smith with a bit more sophistication (putting the moral sentiments into the wealth of nations) and appreciated Marx as more of a critic than a madman. At the end of the day they received generous scholarships to America's elite economic schools where all humanity will be beaten out of them and I'm just the mad Marxist bastard who thinks himself clever for having managed to get the Koch brothers to pay for a few additions to his library, numerous cups of coffee, and a few dinners.

My stint with the Koch brothers revealed what Stephen Colbert, American comedian and host of *The Colbert Report*, taught the United States and what has largely been the case for satire and parody throughout history. They work really well for exposing to those who already know what is really going on. Historically, it is fun to look back and chuckle, but rarely do these works change the times they find themselves within. Although there is something to satire, postnormal times requires something more sophisticated or, at the least, cleverer. As all the great titanic battles of history, literature, and film play out in my head as I try to find a solution to my quandary, one key film comes back to my mind.

Never Look Away is the latest film by writer and director Florian Henckel von Donnersmarck and has taken the film festival circuit by storm similar to his 2006 breakout masterpiece *The Lives of Others*. *Never Look Away* has everything you could ask for from a film, modern art, Nazis, young love, World War II action, and a kooky yet wise old mentor. Allow me to take this film, as its protagonist Kurt takes the horrors he is faced with, and distort the image so that I may not look away, but instead propose a navigation towards a solution. Essentially the film gives us three characters faced with the question of how do you speak to someone trying to kill you?

First, we are introduced to Kurt's father, Johann. The film starts at Germany's finest hour during World War II. Adolf Hitler himself is preparing to make a celebrity appearance in Dresden, the boyhood home of Kurt. Johann is a school teacher, but since he refuses to join the Nazi party, knowing that there is a darker underbelly to the organisation, he is unemployed. Eventually, desperate and faced with the starvation of his family and the potential horrors they could inflict on his sister Elisabeth, he begrudgingly joins the party. Two years later Germany fall to the allies and Johann, now a former member of the Nazi party is 'unhireable' in the then Soviet occupied East Germany. Unable to cope with the tricky cycle of existential crisis resulting from the constant changing of enemy and ally and in deciding which cause he is suppose to dedicate his life, Johann instead takes his own.

On the other hand, we have the character of Professor Seeband, the Nazi medical doctor extraordinaire with the sharpness of Sherlock Holmes and the metrosexuality that leaves Prince looking like a model Puritan. Seeband changes allegiance seemingly with the direction of the wind. As the Nazi's fall, he denies any involvement, even when faced with seemingly irrevocable evidence presented to him. Using favours and good deeds he works his way up into good standing in East German society, but when his former Nazi boss's trail gives hints of revealing his true past, he takes his wife off to the West where the capitalists wont care about his past as long as he can give them a radical new technology or something they can monetise. Seeband is naturally the man responsible for the death of Kurt's beloved Aunt Elisabeth and is in fact the father of Kurt's love interest, Ellie.

While Seeband does a much better job of coping with those who are trying to kill him than Johann, at least that is until Kurt stumbles upon a trail of clues, he hasn't given us a model that we can use. Seeband is a horrible example of the persistence of postnormal lag. Seeband never really stops being this Germanic Hegalian Hero that he makes himself out to be as he becomes the greatest medical doctor to walk the Earth. Thus, he never really stops being a Nazi. This is the man who, without question or much of a second thought, sentenced woman diagnosed with hysteria and special needs persons to execution so as to free up hospital beds for German soldiers. Now, perhaps people change, but this man, years later after completely disavowing his Nazi identity, so concerned with the purity of his Aryan bloodline, made his daughter sterile.

Our last character, Kurt, is faced with this question within the first five minutes and struggles with it on and off until the film's conclusion. Aunt Elisabeth takes young Kurt to the Dresden Museum of Modern Art where a Nazi tour guide denounce modern art as crude, asking those listening to help in reviving a new art for a Germany of National Socialism. Even the Nazis understood the eternal importance of art and its role in preserving and transmitting narratives through time. She delivers the titular line which carries Kurt through the rapid transition of Post-War Germany. The mantra, Never Look Away, follows Kurt as he takes mental snapshots of his life, blurring the most horrific so as to cope, only later to correct them so that they may be projected through his art. Kurt looks on as his Aunt Elisabeth begins acting out, as she is dragged away to the sanatorium, as Dresden is firebombed by allied troops, as his mother is raped by Soviet occupiers, at his father's lifeless body, as he meets the woman of his dreams, as she faces her darkest hour, as he is commissioned to paint the father of the woman he loves who is responsible for his Aunt's death, and all of the ups and downs Kurt sees.

Like his father, he is told to first hate the moderns and to love the propaganda art of the Nazis, then to switch to Socialist Realism, only to be confronted by the scattering styles of the avant-garde in Dusseldorf. Instead of falling behind in such contradictions, or simply jumping from one to the next, he masters each and carries all of them through eventually finding his own unique style in the end. He also watches as the construction of the Berlin Wall is commenced, as if it were another mural, another job. The symbol that stands both as physical barrier and the metaphorically slicing the world which is made into almost a character in such films as *Possession* (1981) and *Suspiria* (2018) is reduced here to another silly work of art.

The last few years have seen a minor proliferation of artist biopics and artist centric films. They all tend to use the drama of the misunderstood artist to help make sense of times of madness, how fitting to the contemporary era. *Never Look Away* takes a different approach. It is not focused on the artist necessarily. Lars von Trier's *The House that Jack Built* (2018) takes the typical art-film approach by using his protagonist, a serial killer, to self-aggrandise von Trier's own controversial portfolio of viewer-discretion-advised films comparing the work of an artist to decomposition in a clever bit of meta narrative. The visually stunning and brilliantly acted film *At Eternity's Gate* (2018) portrays Vincent Van Gogh's descent into ear-severing insanity. Ultimately the piece ends with Van Gogh producing remarkable art, yet its going unnoticed during his life time ran with his inability to discern if he had been attacked, or if his fatal blow was self-inflicted, telling us, the audience, that he was unable to translate his vision to the world, let alone himself. *Never Look Away* gives us something hopefully refreshing, asking us to go to the next level.

Kurt is removed from the art. Obviously talented, his arc is to take this talent and mould it into a visual translation of the things his eyes have seen. Kurt even goes to such extremes as to lie to the press, saying his paintings, which we, the audience, know are inspired by the people closest to him, are simply inspired by random photographs he came across in the newspaper. This detachment, even though artificial gives us a hint at the power of film, especially in postnormal times. Film is an extraordinary medium that can cleverly provide the language needed for confronting those who, whether or not they are aware, are trying to kill us. As this film points out, art lives beyond a given time, movement, or situation. An improvement in the global collection of climate change films from popcorn munching action adventure, to something more cerebral is a wonderful place to start.

It is also necessary to take things from the artist's creation to the real world. Thus, it takes creativity and imagination, beyond the artists, on the part of us all to not simply live within our worldviews (a challenge increasingly no doubt in social media's ability to abide such tendencies), but break beyond them and speak other languages, and here I don't simply mean learning French or Malay! We can begin by ceasing the mindless social media outbursts that generalise the enemy into a hopeless idiot, instead reaching out to them. If logic is the force we make it out to be, I believe it can stand its own in the higher courts of justice. We also have a few terms that we can begin with, for instance, sustainability. In all its various definitions sustainability can be used to both save the planet and destroy it. It is up to us to negotiate how this term is used in the future, determine what decolonising it need endure and what globalising need it be subjected to for the betterment of the planet and all humanity. It is incumbent upon each and every one of us to work to reconstruct Babel's Tower, brick by brick.

ARTS AND LETTERS

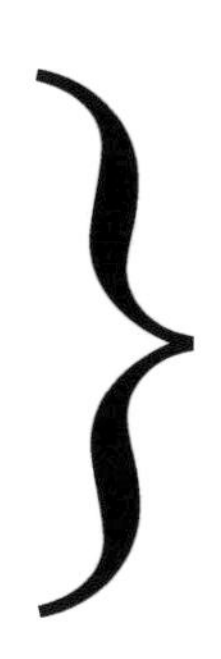

DOUBLE EXPOSURE

Marjorie Allthorpe-Guyton

The Venice Biennale is an irresistible, indigestible moveable feast. Like a Chinese Lazy Susan, the table turns each time, pushed forward by geopolitical shifts. New guests are Ghana, Madagascar, Malaysia and Pakistan, but Algeria and Kazakhstan cancelled and with the dire political chaos in the country, the Venezuelans have not turned up, their pavilion padlocked shut, an austere monument to its architect, the great Venetian modernist, Carlo Scarpa. This again begs the question, is the national pavilion redundant? At a Biennale panel in 2001 the format was described as a 'charming anachronism', prompting the riposte from the British Czech artist Pavel Büchler saying there should be 'pavilions for countries that no longer exist, so that we don't forget what the world is like'. In recent Biennales the national pavilion has come under fire for its separatism, an anomaly in the twenty-first century when borders are fluid and contested, artists are nomads and convergent liberal globalisation may have collapsed into dangerous nationalism and ethnic and religious divisions. This simplistic overview does not take account of the paradoxical and growing polycentricism of international relations. While the power of nation states is threatened by environmental change and by their entanglement in supranational private and corporate networks driven by technology, this does not negate the nation state but repositions it. The sheer scale of the Biennale and the paradoxes of the work of participating artists in the eighty-seven national pavilions, two main exhibitions in the Giardini, the vast Arsenale and in thirty-five offsite churches and palazzi and collateral events, is a paradigm of this seismic shift. For Venice, sinking, in demographic and democratic decline and fed by and consumed by tourism, the Biennale becomes ever emblematic and powerful.

For developing countries, especially in the global south, the survival and building of the nation state is not incompatible with promoting a greater awareness of their national cultural identities, critical needs and democratic aspirations. For them the Biennale offers a singular opportunity. There are now two distinct Biennales, one where the old imperial powers hold on to their elevated places in the Giardini, their shows underpinned by the international art market and another, as yet inchoate and impoverished, where the rest of the world's nations and their artists jostle for spaces and recognition. These so called 'rogue' pavilions which come and go each Biennale cannot be overlooked. If they struggle to challenge the better resourced and established nations which get all the media and art world attention, they can offer a more nuanced view of the potency of contemporary art now.

Kaeka Michael Betero, Daniela Danica Tepes, Kairaken Betio Group
Kiribati Return
Interactive animation and video still 2019

Far west of the Giardini, in a cramped white room In Palazzo Mora, Cannaregio, an animated film has a lone warrior on a tropical beach awkwardly replicate the movements of the viewer opposite who is separated by a low stone wall. Welcome to Paradise! Kiribati, a nation of thirty three remote Pacific islands, first showed at the Biennale in 2017. The survival of its indigenous fishing and farming culture is threatened by rising sea levels and by the ambitions for tourism of its current President. The low cost, unmediated images of artists Kaeka Michael Betero, Daniela Danica Tepes and the Kairaken Betio Group will not draw the crowds, but the plight of Kiribati exemplifies the simultaneous impacts of climate change and of economic globalisation which have differential consequences for all: human and non human. It is this double exposure to multiple socioeconomic stressors, especially technology, urbanisation, changing land use and the consequences for ecosystems and peoples, especially indigenous communities and nations, which is the driving force of much of the work in this Venice Biennale.

Pakistan's inaugural presentation, *Manora Field Notes* by Naiza Khan, is a poetic reflection on place, community and loss. Manora Island, a naval defence outpost off Karachi's harbour, and site of colonial histories and religious diversity, has undergone ecological degradation, architectural destruction and displacement of communities: a microcosm of Pakistan's ongoing industrial and social change and more widely of the global south. Kahn's three part work of archival research includes brass maps and films of her collaborators, local artisans who trundle gaudily lit model boats for sale along the beach, camels loping slowly by. It is a scholarly project, empathetic, but no radical call to arms against prevailing injustice.

Naiza Khan: *Sticky Rice and Other Stories*
Four channel film 2019

Far more affecting is the film of ISUMA, Canada's first indigenous Inuit production company, co-founded by Zacharias Kunuk, Paul Apak Angilirq and Norman Cohn in Igloolik, Nunavit in 1990, and presented, an Inuit first, in the Canadian pavilion. Savagely funny, *One day in the Life of Noah Piugattuk* 2019, recreates an encounter in Baffin Island in 1961 when one family is ordered off their land. For the foot sore *Biennialist* it is well worth the 112 minutes viewing time as a slow unfolding of the misunderstandings and mistranslation in the dialogue between Inuit elder and government enforcer builds a sense of rage and impotency. The ISUMA collective is rare in this Biennale, indigenous artists have only recently found a place on the contemporary art merry-go-round, notably in the vastly ambitious and relatively well funded Documenta 13, the quinquennial exhibition In Kassel, Germany, directed by the star curator Carolyn Christov-Bakargiev in 2012. This was ahead of the curve; in an unprecedented move, indigenous peoples are now being recognised as part of the solution to the global current crises. The groundbreaking 2019 UN Global Assessment Report on species extinction urges 'explicit consideration of the views, perspectives and rights of indigenous Peoples and Local Communities, their knowledge and understanding of large regions and ecosystems, and their desired future development pathways.'

ISUMA: One Day in the Life of Noah Piugattuk
4K digital video 2019

No artist better articulates the wisdom of indigenous peoples and what the theorist Bruno Latour describes as 'The Great Divide' between, object, culture and nature, than Jimmie Durham who was awarded the Biennale Golden Lion for Lifetime Achievement. Durham was nominated by the Biennale's American curator, Ralph Rugoff, Director of the Hayward Gallery, London, who commends him 'for making art that is at once critical, humorous and profoundly humanistic.... often accompanied by texts that drolly but incisively comment on Eurocentric views and prejudices. Insistently invoking the limits of Western rationalism and the futility of violence'. Durham is represented in the Giardini exhibition by a large block of Serpentine stone and a text which baldly states its long 'serpentine' route to Venice: 'The stone was quarried in northeast India, close to the border with Myanmar. Local people, 'tribals' of the Loi, Kuki and Naga, are the quarry workers. They are not the owners, however....' In the Arsenale, Durham shows his signature sculptures 'illegal combinations with rejected objects': solitary skeletal beasts sensitively created from manmade waste. It is a pity they are few and packed in too tight; but maybe this is the point. The viewer is left wanting more of these magnificent augurs of a depleted world.

Jimmie Durham: *Musk Ox* 2017

Alongside legions of academic texts on the theme 'Contemporary Art and the Politics of Ecology', Durham writes, in the special issue of *Third Text*, January 2013, 'Against Internationalism'. 'What' he asks, 'for example, if we had a world law against the buying and selling of land? If, as so many stateless peoples say, the earth is the earth and not a commodity?' Durham's award is long overdue. It is the paradox of this Biennale, a site of privilege and excess, that Jimmie Durham is its lodestar. He carries the flame of the artist Gustav Metzger (1926-2017) who long addressed the planetary dysphoria which charges this Biennale. Metzger's project *Extinction Marathon; Visions of the Future* 2014 and worldwide day of action *Remember Nature* 2015 were the immediate precursor of the worldwide Extinction Rebellion movement. Both Durham and Metzger attack the privileged and autonomous positioning of art, as Metzger said 'there is no choice but to follow the path of ethics into aesthetics. We live in societies suffocating in waste'. This path is passionately pursued by another elder, the artist and founder of *Third Text*, Rasheed Araeen, whose seminal *Art Beyond Art, Ecoaesthetics: A Manifesto for the 21st Century* 2010 argues for a collective art practice that integrates with collective struggle. Viscerally and visually compelling is the collaborative work of Soham Gupta at the entrance to the Arsenale. Gupta's harrowing portrait photographs from the series *Angst* 2013-17 are intimate encounters with the marginal poor and abused in Kolkata which give the subjects some sense of agency in the world.

But the danger is that Araeen's clarion call is met by the kind of literal, virtue signalling exemplified by the title of the US *Brooklyn Rails'* collateral exhibition in Canneregio, 'Artists Need to Create on the Same Scale that Society Has the Capacity to Destroy', lifted from a neon work by Lauren Bon. While in the Arsenale, *Crochet Coral Reef* by the sisters Christine and Margaret Wertheim may be the work of 10,000 crotchetiers making startlingly beautiful simulacra of rainbow corals out of beads and threads, described as, 'an elegant alliance of science and art', they are nothing more. More conceptually subtle and thought provoking is the work of Argentian Tomás Saraceno, his *Aero(s) cene*, sound and cloud forms suspended in the Arsenale marina, bridges scientific disciplines to envisage a time when we can claim back the air, now traversed by 100,000 polluting planes every day. His strange and witty collaborative project in the Giardini, *Spider/Web Pavilion 7, Oracle Readings, Arachnomancy, Synanthropic Futures: At-tension to invertebrate rights!* 2019, confronts the viewer in a blacked out cell with the astounding strength and complexity of spider webs spanning the space, seemingly with no support.

Poster for Brooklyn Rail Exhibition

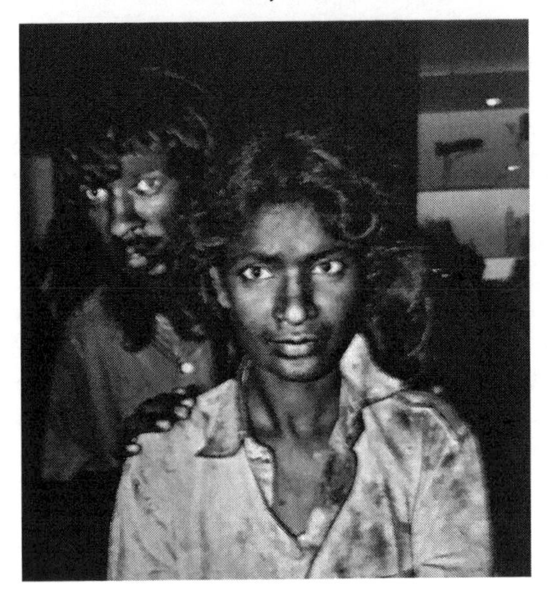

Soham Gupta: *Untitled*, Angst series 2013-17
Pigment print

A poignant lament to the last days of an exhausted planet, and one of the most resource intensive and least funded of all national pavilions, is Lithuania's 60 minute micro opera installation *Sun and Sea (Marina)*, winner of the Golden Lion for best national pavilion, beating the favourite, the highly pleasurable, weird and enchanting song of the sea, *Deep See Blue Surrounding You,* by Laure Prouvost in the French pavilion. The Lithuanian Nida Art Colony of Vilnius Academy of Arts managed to raise $40,000 through crowd funding from over a thousand backers to help pay for costs, not including the rent of Magazzino 42, Marina Militare. In the austere redundant space, over thirty volunteer performers sing and soak up a fake sun on a beach made of 35 tonnes of Lithuanian sand, overlooked by the audience on a mezzanine above: 'my eyelids are heavy, my head is dizzy, light and empty body, there's no water left in the bottle…too much sun.. Empty snail homes, swollen seaweed, fish remains, and all kinds of shells'. One performer posts 'lying here on the beach wearing swimsuits made in the factories of China, is this not a parody of the Silk Road? The historic Silk Road started and ended in Venice. With its thousands of wealthy collectors flying in to view, and sometimes even to buy … is the Biennale itself not such a parody?'

This Brechtian contemporary parable is a collaboration between a theatre director, Rugilė Bardžiukaitė, a playwright, Vaiva Grainytė and a composer LIna Lapelytė. Bardžiukaite's film, *Acid Forest* 2018, won the Swatch Art Peace Hotel Award at the Locarno Film Festival. *Sun and Sea,* first performed in Vilnius in 2017, is curated by Lucia Pietroiusti, Curator of General Ecology and Live Art at the Serpentine Gallery, London. The poetic resonance, topicality and accessibility of this multi art form work which aims to address 'ecological issues and the Anthropocene', attracts the viewer away from the more orthodox contemporary art works of painting, sculpture, and non-narrative film and video, such as the chilling two channel film of Larissa Sansour, *Heirloom and In Vitro,* in the Danish pavilion. A mother and daughter sustain life in a bleak concrete bunker after a cataclysmic disaster when a tidal wave of viscous black oil drowns a city, Bethlehem. In grainy black and white, the work is a visually arresting black and white reflection on ruin, loss and trauma in contested lands. This is far removed from the *chi-fi* banalities of more literal contemporary art re-presenting contemporary ravages, but it leaves the art worn viewer jaded and seeking solace elsewhere. Even *48 War Movies*, 2019, a dense single channel video of layered images with deafening soundtrack in the Arsenale by Christian Marclay, whose *The Clock* 2010, must be one of the most watched contemporary art works ever, struggles to hold the attention of Biennale crowds seeking out more immersive experience and live performance.

This is delivered in spades by the Brazilian pavilion where *Swinguerra*, the film of Barbara Wagner and Benjamin de Burca, enthrals through its sheer transgressive cheek. Strong young dancers, some transgender, throw their lithe and ample bodies through the moves of the popular subversive and highly politically charged *swingueira* movement of north east Brazil. *Swinguerra*'s life and optimism casts even Lithuania's much feted *Sun and Sea (Marina)* as a work of a pallid ecoaesthetics born of relative affluence and, like the confrontational avantgarde of the past, ultimately a futile gesture.

The artist who is renowned for her exceptional grip on the power of science and technology to change the world, for good or bad, is Hito Steyerl whose Biennale works are repellent gaudy videos made by artificial intelligence (AI) video processing. *Leonardo's submarine* 2019 in the Giardini exhibition metaphorically turns the narrative of Leonardo's invention of a proto submarine to throw sharp light on an Italian armaments company which has changed its name to Leonardo S.p.A. Steyerl leads the expansion of high tech in contemporary art of which the Biennale offers plenty. Given the luxury of time to track down all the works, the viewer can don headsets and hallucinate with ectoplasmic forms in Dominique Gonzalez-Foerster's *Endodrome* 2019, or indulge a sensory virtual encounter with melting ice caps, and the legendary Marina Abramović, in the palazzo Ca' Rezzonico.

Steyerl is a highly bankable international artist; and the financial bedrock of the Biennale is now the international art market. At the 2001 Biennale, The Pavilion of Turkey had no government funding and was hosted in the Italian pavilion. It was independently curated by Beral Madra who argued: 'who is supporting this Biennale? ...When I go to the national pavilions, I don't see (just) nation. I see technology, I see multinational companies. I see a kind of cultural entertainment. I see promotion. So all this is related to multinational capitalism...business, as *art* business, is absent in these unrepresented countries.' In 2019 has anything changed? While most of the 83 artists selected for this Biennale by Ralph Rugoff were born outside the US, a quarter are based in the US. Many are represented by the big international dealers such as White Cube, Victoria Miro, Gagosian, Spruth Magers, but Rugoff succeeds in breaking the art fair mould. He lays out a cogent critical platform for his exhibition in the Biennale 'short guide' and meets the paradoxes of Venice head on. The art is given more space, especially in the vast Arsenale building which has temporary plywood walls, a counter to the pristine white of the art fair booth. Rugoff has reduced the number of artists (of which more than half are women), chosen only living artists and invited them to show in both the main venues but with distinctly different works: *Proposition A* in the Arsenale, *Proposition B* in the Giardini. It is a credit to Rugoff's concern for the 600,000 expected visitors over the six months of the Biennale that they should see the artists in greater depth, to be able to read their multiplicity, ambivalence and contradictions. He quotes Jimmie Durham: 'I don't want to say what I would say, because then it becomes me talking through the piece. I want to see if I can make the object have a conversation with whoever's looking'.

Rugoff's knowing ersatz title 'May You Live in Interesting Times', is not so much a theme like the earlier Biennales 'The Environment' 1976, 'Future, Present, Past' 1997, but a *dimension*, like the 'Plateau of Humankind' set by the legendary curator of the 2001 Biennale, the late Harald Szeeman: 'why the hell shouldn't we try, even if it sounds not so good to some people, to (formulate) a direction which goes in all directions?' But all roads do not lead to Rome. In the existential gloom of 2019, when destructive alienation, of peoples and individuals, is at a crisis and ecosystems are degraded towards extinction, can art keep faith with its social function? And fulfill Marxist Ernst Fischer's dictum that art must 'show the world as changeable. And help to change it?' Fischer's prescient *The Necessity of Art* 1959, sought to declaim the contradictions of outworn capitalism: 'as man becomes more and more capable of mastering and transforming nature and the entire word around him, so does he confront himself surrounded by objects which are the product of his activity yet which have the tendency to grow beyond his control and to become more and more powerful in their own right'. Rugoff acknowledges the urgency of contemporary conditions that fuels the Biennale exhibitions, but does not claim for art more than its ability to guide, to 'illuminate aspects of our social relations and psyches'.

Barca Nostra 2018–19
Shipwreck 18th April 2015

Some have aimed for much greater agency, none more so than Swiss Icelandic Christoph Büchel 's project *Barca Nostra 2018-2019, Shipwreck 18th April 2015*. With Sicilian and Italian collaborators, Büchel had the rusted hull of the fishing boat in which 800 or more migrants drowned off the island of Lampadusa, Italy, transported from Sicily, where the bodies were recovered, and moored right alongside the Arsenale café, outraging critics. Büchel sees Venice as 'a city based on migration that feeds the machine of its own destruction through mass tourism' and claims the boat 'as a vehicle of significant, socio-political, ethical, and historical importance'. In the 2015 Biennale he set up a mosque in a former catholic church, the official Iceland pavilion; it was closed after two weeks. In 2017 he constructed a facsimile of the Calais Jungle site in a contemporary art gallery in Ghent. All works intended to probe and provoke. Does *Barca Nostra* fail as art because it crosses a crucial line and becomes an unacceptable and distasteful confusion of art and reality? As Büchel ruled out any visible notice of its history on site, the boat's identity as a tragic sarcophagus of the recent dead eludes the viewer. Those two outdated fixtures of the Biennale, the bald performance artists Eva and Adele, in their trademark pink frocks, posed for selfies in front. Unlike the powerful 2018 film *Styx* by Germany director Wolfgang Fischer about the moral crisis presented by migration, *Barca Nostra* is not an art work. A deeply disturbing and grim presence at such a cultural feast, it is an activist's provocation, an act of placement. It is an effective enforcer of the collective mind to focus on continual moral crises; which the roving Biennialist, and the media, can choose to forget and from which much of the art, however well intentioned, is an aesthetic diversion.

PLASTIC POLLUTION

Greenpeace

An estimated 12.7 million tons of plastic – everything from bottles and bags to microbeads – end up in our oceans each year. That's a truck load of rubbish a minute.

Travelling on ocean currents this plastic is now turning up in every corner of our planet – from Cornish beaches, to uninhabited Pacific islands. It is even being found trapped in Arctic ice.

It is hardly surprising then that our oceans are slowly turning into a plastic soup: the effects on marine life are chilling. Big pieces of plastic are choking and entangling turtles and seabirds and tiny pieces are clogging the stomachs of creatures who mistake it for food, from tiny zooplankton to whales. Plastic is now entering every level of the ocean food chain and even ending up in the seafood on our plates.

The oceans produce half of our oxygen and food for a billion people. And because they soak up huge amounts of carbon dioxide, they're also one of our best defences against climate change. Our fate is bound to the fate of our oceans. If they don't make it, we don't either. Greenpeace is campaigning to end the flow of plastics into our oceans. These images illustrate why we need to act now.

Coral Bleaching in the Maldives
Credit: Uli Kunz/Greenpeace

Plastic caught up in the bleached corals around the Addu Atoll in the Maldives. Algae living in the corals are released as a stress reaction caused by warmer water temperatures, thus turning the coral white. The corals can only revitalise once lower water temperatures return. If this does not happen, they die. In February and March of 2016, water temperatures of 32 degrees celsius were measured over several days around the Addu Atoll, possibly a result of El Nino.

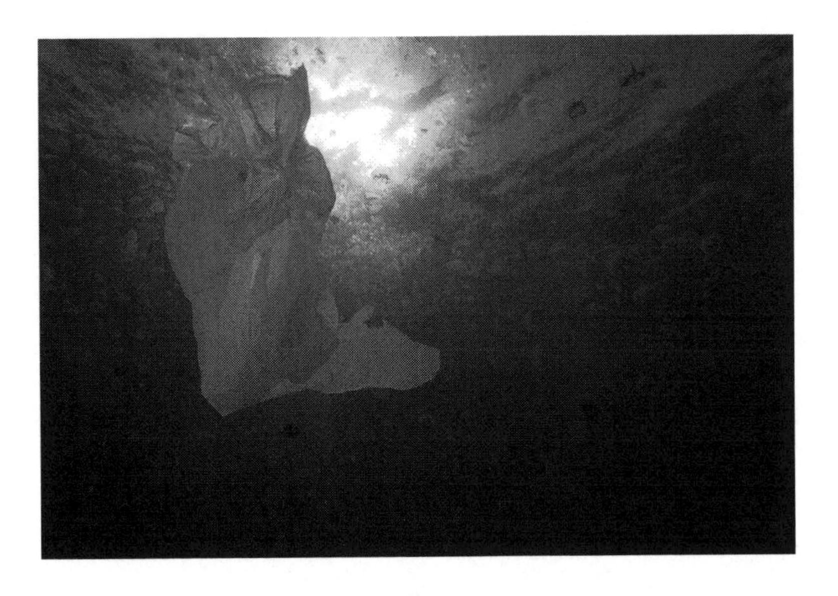

Plastic Bag in Water – Red Sea Coastal Development in Egypt – 2006
Credit: Marco Care/Greenpeace

A plastic bag floating in the waters of Elphinston Reef, near Marsa Alam, is easily mistaken for a jellyfish. Garbage such as this snares and chokes marine animals which misjudge its shimmering appeal. Elphinston Reef is a spectacular reef and one of the most beautiful sites in the region. Crystal clear waters and unique coral reefs have made the Red Sea one of the world's prime diving destinations. Yet these reefs are threatened by problems such as overfishing, pollution and uncontrolled coastal development.

Plastic Waste at Manila Bay Beaches
Credit: Daniel Muller / Greenpeace

Bird flies over water that is polluted with plastic waste in Manila.

Freedom Island Waste Clean-up and Brand Audit in the Philippines
Credit: Daniel Muller / Greenpeace

Greenpeace together with the #breakfreefromplastic coalition conduct a beach cleanup activity and brand audit on Freedom Island, Parañaque City, Metro Manila, Philippines. The activity aims to name the brands most responsible for the plastic pollution happening in our oceans.

Freedom island is an ecotourism area which contains a mangrove forest and swamps providing a habitat for many migratory bird species from different countries such as China, Japan and Siberia.

Blue Footed Bobby with Plastic Waste in Peru
Credit: Robert Marc Lehmann/Greenpeace

Seabirds/Blue Footed Bobbies sitting in between plastic waste on the beach (Sula nebouxii). Isla Lobos de Tierra, Peru.

Plastic Waste on Coral Reef in the Indian Ocean
Credit: Wolf Wichmann / Greenpeace

Underwater image of plastic waste in a coral reef on Abd al Kuri, a rocky island in the Indian Ocean.

Plastic Bottle in the North Sea Credit: Will Rose / Greenpeace

Plastic bottle floating in the sea in the Firth of Forth, Scotland. Up to 12.7 million tonnes of plastic waste enters the oceans every year.

Garbage on the Beach at Addu Atoll in Maldives
Credit: Uli Kunz/Greenpeace

Piles of garbage on the beach of Gan Island of the Addu-Atoll, Maldives. Plastics, packaging and textiles make up a large part of the rubbish.

Dead Fish and Plastic on Beach in Mexico
Credit: Greenpeace

In occasion of Earth Day, volunteers in Mexico joined together to collect evidence of single-use plastics belonging to the world's largest corporations. Greenpeace is asking that they reduce their production of single-use plastics.

SHORT STORY

LOOKING FOR WATER

Uzma Ali

We walked for days. I had placed the child with all her apparel on a donkey. They were the only livings things left in the village. She looked serene, in her red robes and priestly breast plate, despite the fact that our village had just been obliterated moments before.

I watched the village from a foothill outside as it was ransacked by The Marauders. Large soldiers, broad and armoured, tossed people aside as if they were feathers, charging into every house to see what they could find. I wasn't sure if they were looking for treasure or water stores, but they destroyed everything, setting fire to almost every dwelling. The fighting and screaming kicked up a fine brown dust in the dry air, and a haze hung in the pale yellow light as far as the foothills in which I was hiding.

A group of villagers began to form a crowd outside the temple. I couldn't see exactly what they were doing, but whatever it was, they seemed to provide a significant enough resistance, because The Marauders movements slowed. But nevertheless, it wasn't enough. It looked to me as if eventually all the villagers were killed, and then a couple of The Maruaders went straight into the temple, and left again.

The image of the slain villagers stung deeply. It reminded me of how my family had died many years previously. When all of the Marauding hordes had left, I scrambled down from the foothills, and surveyed the scene. The air was dry, the walls of the circular mud-huts were arid, and dead bodies were scattered everywhere. There was no blood or scars, The Marauders were well known for using electrocution as a means of combat. Some of my closest friends were among the dead, but I had no time for sentimentality, I needed to think quickly about how I could survive.

I entered the temple, gaining respite from the sun and horror outside, relieved to be away from the carnage. The stone cool of the building enveloped me, and I walked slowly towards the holy of holies, in a way I never have imagined I would have the courage to. Before I got to the inner sanctum, I felt a tugging at my clothes, and saw the small girl, dressed in finery, at my feet. She had lived in the holy of holies and was one of those responsible for finding water for our village.

Several years ago, explorers from our federation had set out on foot looking for water, and more often than not, never returned. Killed by vigilantes or presumably having died of thirst. These men had wanted to be heroes, but all they had become were tragedies in a collective folk memory.

Then the child had been born, with the mark of the skill on the back of her left hand. Three raised moles in a perfect line. People had watched desperately, as the small baby twitched in her mother's arms, hopeful that her movements would somehow communicate where water could be found. A small girl from the village, called Hafisa, screeched the first time she laid eyes on the child. She could hear what the baby was saying, and she had said that there was a stream of water, under the ancient rocks just outside the village boundary. Men set to work to move the ancient stellai and discovered, at the base, a fine layer of moss.

The moss was full of water and when chewed, would quench our thirst. When the baby became a child, she was able to point out where the spirits had told her water was located, and invariably a water source would be found, usually though some desert plant or nearly dried spring. Through the intuition of The Oracle the village survived. When the village got lucky, and the oracle's powers were at their most strong, she would direct them to a cave or underground stream that contained enough water for drinking, washing and storage. Intermittently, the child would catch a fever and shriek in a febrile stupor for days. These fits always preceded rain, and the village would collect as much rainwater as possible, dancing in the rain for days. But these joyous occasions were rare. The rain was infrequent, and drought was far more common than days of plenty.

Then The Marauders came. For what, I wasn't certain, but the health and stealth of the fighters betrayed the fact that they had been fed and watered in a way our village dwellers could not even dream of. The Marauders viewed us as primitive and disposable. Perhaps they had come for temple treasure, maybe knowing that our oracle had the power of divining water sources.

And now, at the child's bidding I walked through the wilderness, in the hope we would find water. We walked through mountain paths, and dried river beds, and every so often the child would point frantically. There had been an intentional effort on the part of the temple officiates not to teach the child language. It was thought that this would sharpen the child's divinatory powers. So the child was only able to point, screech, and sometimes utter a few words.

Following the instructions of the child, the Oracle, I would discover a small shrub or mollusc that would contain water, and we would share it. The child's power lay in being able to differentiate between poisonous plants and creatures, and those that were safe to eat. Regardless of generations of study and collective wisdom, there had never previously been anyone in my village who was able to consistently direct people towards safe drinking water. We were entirely reliant on the chance birth of a diviner.

My face was parched and tired, and my feet sore and calloused, but the child seemed determined to keep moving. When I placed her on the donkey, she would bundle her red robes in her lap and scream *there, there,* pointing straight ahead. I had no choice, but to lead the donkey by its tether. We had been moving for about four days, setting up a humble camp in the shadow of rocks and boulders, surviving on the smallest of shrubs and insects, and a tiny morsel of bread I had found in the village. I feared that we would come across a group of Marauders or Counters, and they would kill me and imprison the child. There was a brooding that grew inside me; *aren't we moving in the direction of Counter civilisation.*

Because to move due East was to move towards the more densely populated Counter areas, and there we were sure to be enslaved at best, and killed at worst. We were simple villagers. Our ways were mocked, and we were considered little more than animals in the eyes of Counters.

But still the child directed me towards the cities of the Counters. A fate that I would rather have avoided. We had now come to the edge of the salt flats, a level and pristine white, containing salty pools of water I couldn't drink. The sounds of nature, that were intermittent amongst the salt terrain, were now interrupted by a distant humming, and now and then a dramatic clanging. The noise of metal on metal, the buzz of electricity. I thought of the slave populations that were known to keep those cities functioning. I prayed to the spirits of the oracle that I would be able to escape such a destiny.

Still the child wanted to move closer to the cities. Still I pleaded, but the child knew that she exerted an almost total control on me, and I would listen obediently to her instructions. How else would I survive? Without her, I would die of dehydration, and my body would become a dry skeleton like so many others we had passed on our travels.

That way, that way the child said as she pointed. Irritation at the child's assumption of total authority and her inability to speak more than three words at a time, crept over me. I began to suspect that instead of protecting me, she was leading me deep into enemy territory.

On the horizon of the salty plains, there was a thin line of grey, and I knew that this was the line of the city walls. The city walls of the Counter-Revolutionaries, who had been more brutal in the quashing of insurgency than any other known civilisation. Their brutality ensured their survival. The thought of approaching the citadel made me feel ill. I had already escaped death once through sheer luck, I didn't want to look death in the face ever again.

Forward, forward the child said, and I reluctantly pulled on the tether of the donkey, and led the donkey and child towards enemy territory.

The sounds to which I was not accustomed began to fill the air. The creak of salt plains became the whirring of vehicles and the rhythmic tones of industry.

I was certain we were being followed. Counters were known to release colonies of mechanical insects outside of their city walls. This was their major protection against Marauders and other foe. Their expertise in technology and resulting wealth was jealously guarded, and they thought nothing of killing anyone who was several miles from their city walls.

My fears came to fruition when I saw a tiny beetle follow us, jet black against the whiteness of the flats, and when I bent down to look closer I saw metal pivots in its legs and pincers, and a synthetic glassiness in its wings.

The Oracle had led me to my death. For a moment I was expecting a whole colony of flesh eating spiders to emerge out of the ground and start sinking their poisoned pincers into my body. I started trembling, and feeling faint in the same way that I had when I saw the Marauders sack the village.

Forward, forward the oracle screeched. At this point I wanted to do nothing more than slap the child's face. Perhaps all the times she had directed me to water had been coincidence and now her childish ramblings had brought me to death.

A dense silvery cloud seemed to float above the city walls in the distance, and the silvery cloud began to move towards us. As it came closer, I realised that it was a fleet of air zeppelins, with the Crescent symbol of the counters emblazoned on them in red.

My body clenched, and awaited sniper fire. I cursed The Oracle under my breath, and hoped that the spirits would show me safe passage to the next world.

They mix air, shouted The Oracle.

The zeppelins were overhead now, and hung in the sky in an expectant swarm. Then one by one they burst into orange flames, and as each zeppelin turned into a floating fireball, a loud bang moved through the air, like a choreographed thunder storm. I kept wondering how I was still alive.

Then there was stillness, and I felt a drop of warm liquid fall on my face. And another, and another, and another.

Mixing air makes water, said The Oracle.

SHORT STORY

THE CHILDREN ARE READY

Hafsa Abdurrahman (aged 8)

It was the day that could change us forever. The sun slowly rising up into the dark sky. At 8am everyone arose. Some of the children didn't put on their uniforms. They ate their breakfasts as normal but they didn't pack their school bags. Their calendars were all set at day zero. This was the start of it. They were ready.

It was rush hour and the streets of London were busy as people bustled and ran around, going about their business, catching their late trains or just late as they ran to catch their trains. The streets soon became empty. It was unusual. London was never this quiet. Nothing happened. It only took a few minutes and suddenly they were everywhere. Children were gathering. They were ready.

Everywhere. There were children everywhere. Holding big banners up to the blazing sun. They were shouting the words 'Save Our Future'. Some were angry. Some were crying. They wanted things to change. They came in cars, vans, buses, trains and even walked. They travelled far in any way they could to get there. They had arrived. They were ready.

There were police helicopters swarming in the skies like bees. There used to be bees. Didn't they make this kind of sound? The children couldn't remember. They walked towards 10 Downing Street and juddered to a halt outside. One of the police helicopters was so low it deafened them with a sound like the mighty deafening roar of a tiger. There used to be tigers but the children couldn't remember them.

It wasn't the sound of a helicopter. It was a tiger. It leapt out of its hiding place. The children ran. They got home any way they could and shut their doors and had their lunch. But they still didn't put on their uniforms. They were ready.

The Prime Minister wasn't eating her lunch. She was inside 10 Downing Street, still wearing her nightclothes. She had tired-looking eyes. No one would know, she thought. But the children knew. We're not done. She heard them say. What about our future? She heard them say. This question will never end, she heard them say. We are ready. She heard them say.

SCREAMING OLIVES

Farid Bitar

Beyond the evergreen
The settler is mean
The olive tree glimpsed
The oblivious coming
Carrying a tank of petrol
And an M-16 hanging from his shoulder

The branches started waving hurriedly
Then the olives joined in, screaming
Like a chorus in a symphony
Call our owners. Call our owners
But no one comes
Wonder why?

Is it because the soldiers are busy
Sniping the farmers in the harvest
The tree started yelling at the man.
Get out of my existence
Every time you come, I know I'm gone
Stop killing my babies
Stop suffocating my insides
Stop your savageness.

Emmett Till lynched in 1955
Way down America south
Ali on Grill, the settler taunted
Burned in Duma, West Bank, Palestine, in 2015
Mississippi still.

THE HOUSE IS ON FIRE

Helen Moore

After Greta Thunberg

God's own property, some call it,
for others it was built by Allah, Jah, or Rainbow Serpent;
yet most agree it's priceless and that for sheer scale
and beauty it surpasses any maharajah's palace,
its treasures infinitely more splendid
than we could ever fashion.

And this house is ancient –
4.5 billion years by some calculations;
it should be listed
top for conservation due to its uniqueness.

Nowhere else we know of in the Cosmos –
this living, breathing home
extending hospitality to every guest who comes
from spore, seed, egg or womb.

Fire! Fire! the girl cries at the gates of the law-makers,
as she finds its blue and white ceiling
is rapidly overheating.

Who heeds the young?
Having made themselves too comfortable
some men pretend not to hear.

Other people claim: *It's Divine will!*
We should all pray for salvation.

Yet others wear that tattered coat of fantasy
in which they seek disguise:
If it gets too bad in here, we'll leave this old house!

But where can they go?
This is our home and there are flames consuming its rafters.

REVIEWS

GoT: A CLIMATE ALLEGORY?

Yovanka Paquete Perdigao

The finale of the eighth season of *Game of Thrones* (GoT) ended on a sour note for many fans but provided a wave of relief for those who couldn't take the relentless 24/7 #GoT Twitter speculation any more. It would not be hyperbolic to say the show has proven make or break for relationships, divided families, caused friends to fall out, and sparked tension among work colleagues. Loathe it or love it, for more than eight years our weekends have been spent in anticipation of a new battle, a sensational family secret, another disappointing death. Some episodes remain iconic, etched on to our collective consciousnesses. Let's just say I will forever feel a shiver down my spine when the 'Red Wedding' is brought up and 'The Bells' will always elicit multiple kissing teeth sounds.

We started episode one, series one, full of hope. We religiously followed the House of Stark of Winterfell, the dominant house of the North of Westeros, which is the mythical land where GoT takes place. The series dutifully opened where the first book by George R R Martin's *A Song of Ice and Fire* began, although key to understanding the series and the book is the devastating event referred throughout as 'Robert's Rebellion'. It is this event that is so central to the current configuration of the Seven Kingdoms, the Royal Houses in Westeros, and our characters' ascent to their current positions.

Years before *A Song of Ice and Fire*, the Seven Kingdoms were governed by the Targaryen dynasty, who sat on the Iron Throne that ruled all other royal houses. Foreigners from Old Valyria, the Targaryen were forced to flee a catastrophe towards Westeros. Upon finding sanctuary they battled the different kingdoms to rise to the top, thanks to their relationship with dragons. All the other royal houses would now kneel before the Targaryens and act as vassals of the crown. However, up North, Lord Rickard Stark harboured ambitions for the House of Stark. His plan was to marry his son

Brandon to Catelyn Tully of Riverun and his daughter, Lyanna, to Lord of Storm's End Robert Baratheon, in a bid to cement alliances between the House of Stark and the rest of the Southern houses. His other son Eddard, (Ned), was extremely close to his siblings and grew up alongside Baratheon with whom he enjoyed a deep friendship. But the fate of the young couples was sealed when Lyanna Stark disappeared, shortly before the wedding, believed to have been kidnapped and raped by Rhaegar Targaryen. The eldest, Brandon Stark, set out to seek justice before being arrested by King Aerys II, father to Rhaegar and ruler of Iron Throne. Brandon's father, Rickard, was also caught and both father and son executed for treason. Aerys II who had been slowly descending into madness, attempted to arrest Ned Stark and Robert Baratheon to quell any challenges to his rule, provoking the ire of the other Houses, and igniting Robert's Rebellion.

Ned ends up marrying his brother's fiancée, Catelyn, upholding the Stark's promise to the Tullys, but immediately rides off to join the battle in King's Landing. There, at the centre of power in Westeros, with Robert and the remaining houses, a terrible and long war begins. Ser Jaime of House Lannister put an end to the war when he slayed King Aerys II, by then referred to as the Mad King. The remaining Targaryens are brutally tortured, raped and killed except for Aerys' youngest children, a boy and a girl who managed to escape King's Landing. Robert is crowned King and is betrothed to Cersei Lannister, sister of Ser Jaime. Ned, having witnessed too much gratuitous horror, disagreed with the brutal murders of the Targaryens, and set out to find his sister Lyann. She is discovered in a secluded tower, just as she is about to die of childbirth, and reveals her love for Rhaegar, explaining that the two secretly married before the war. In her final moments she pleaded for her brother Ned to protect her baby Aegon. Fearful for him as he is the last remaining Targaryen and, as Rhaegar's son, is rightful heir to the Iron throne, Lyanna demands Ned promise her he will keep her son's true identity a secret. Ned obliges, and raises the baby as Jon Snow, a bastard fathered by him during Robert's Rebellion, keeping Jon's identity a secret from everyone including his wife.

Both the book and television series begin almost fifteen years after these events, with King Robert offering the Handship and marriage between his heir Joffrey and Ned's daughter Sansa. Through a series of events, Ned realises that the Lannisters have been plotting for control of King's Landing and Robert's children including his heir Joffrey are actually the result of an incestuous relationship between Queen Cersei and her brother Ser Jaime. Robert died under mysterious circumstances, Ned is executed for treason and Sansa held captive in King's Landing. This prompts a rebellion from the House of Starks who soon find themselves betrayed by their own allies, leaving the bastard Jon Snow, Sansa, Arya, Bran, the only surviving children of the House of Stark.

Throughout the story, all four children embark on different journeys that take them across Westeros whilst also following the last Targaryen, Daenerys, who is slowly rising to claim back the Iron Throne. Daenerys marries a Dothraki fighter and quickly becomes a widow with no children but three dragons. She goes on to make a name for herself as 'Mother of Dragons, Breaker of Chains' by punishing slave masters and freeing slaves. She amasses a powerful army of Unsullied and Dothraki fighters alongside three fully grown dragons. Over the years she becomes driven to conquer Westeros and, as the last Targaryen, believes she is the rightful heir to the Iron Throne occupied by the Lannisters.

Meanwhile Sansa tries her best to survive despite being used as a pawn by the many men she encounters, Arya flees to become a skilled assassin to avenge the deaths of her parents and siblings, Bran, now crippled, enters into a spiritual journey to discover his true self. Jon who has always felt rejected and treated as an outsider for being a bastard, exiles himself to join the Night Kings' watch, a military order, which guards the Wall that keeps the Seven Kingdom separated from the wildling and White Walkers.

Legend has it that when the first humans appeared in the World, they started a war with the Children of the Forest, the natives of Westeros. In desperation, the latter captured a man and turned him into a White Walker, the first of its kind, who they named the Night King. Raising an army they sought to protect themselves but the White Walkers broke free and began to terrorise all living creatures in Westeros. The Children of the Forest and humans formed a pact and fought alongside each other to defeat the White Walkers, chasing them far north of Westeros. To prevent them from coming back, the Children and humans created the Wall, a massive fortification standing seven hundred feet high and stretching from one side of the continent to the other and infused with magic preventing the White Walkers from crossing. The Night's Watch was created to protect the Wall but by the time Jon arrives, the White Walkers are nothing but a myth and the Wall just a separation between Westeros and the few men who live outside its protection, called Wildlings, as well as odd creatures here and there.

However throughout the story, it is revealed that White Walkers are far from myth. In fact they are a real threat, slowly advancing to Westeros unbeknownst to our ensemble of characters. Jon gets a first taste when he finds himself amongst the Wildlings and witnesses the White Walkers, tall and white humanoid figures that are hell-bent on killing all living life and possessing the power to turn the dead into walking zombies. Having witnessed their capacity for destruction, Jon's mission is to convince the rest of his world, Night Watchers, the remaining Starks, Daenerys, the Lannisters, and all the other houses, that if they do not act soon, their world as they know it will be extinct.

Extinction is what brings us to this essay, an apt word to describe the threat posed by the White Walkers, and eerily similar to us on the other side of the books and the television screen. GoT is an elaborate and sombre allegory for climate change and we, citizens of planet earth are the people of Westeros, distracted with fighting each other, whilst a silent but deadly threat rises every day. Except our White Walkers are not Zombies but represent the destruction of our planet by our own greed.

Although the creator of *Game of Thrones*, George R R Martin initially refuted this allegory, he has come to embrace its truth as the parallels between his world and our world are uncanny, to the point that GoT sometimes stops being a mere fantasy story. When we are introduced to the creation of Westeros and the Wall, we are confronted with scenes that resonate with our own world: the First Men declaring war on the natives, the Children of the Forest. This is nothing but an old tale of colonialism. The First Men are the equivalent of the Western White invaders who 'discovered' the rest of the world, men who were emboldened by their guns and germs, descending on parts of the world that had thriving civilisations, only to impose their arbitrary rules, ideas, and ways of living. The Children of the Forest are reminiscent of the many indigenous people across the Americas, Africa, Asia, Pacific and their connection to nature is not a coincidence. The First Men appear in George R R Martin's world to conquer the land, and by doing so, destroy the natural habitat of the Children of the Forest. One very obvious example springs to mind, the continuous destruction of the Amazon forest in Latin America, that has primarily been driven by white settlers who in the pursuit of profit, have not only upset the delicate eco-balance of the Amazon but killed, stalked, and displaced animals and Amerindians who had lived in these forests for centuries. Or the Native Americans who welcomed the first white settlers with open arms, only to be slaughtered mercilessly and their history whitewashed by a turkey eating ritual on Thanksgiving. There are countless more illustrations in other parts of the world, but unlike the Children of the Forest, the indigenous people across our world don't have the same magic as the former. What they did have, however, was the ancestral knowledge and profound respect for the planet we all share, hence why many coexisted peacefully with Nature until the Western man appeared.

Although Westeros civilisation did not start with the Ice Age like ours, it was born out of another natural disaster, from the ashes of Old Valyria. Once upon a time, the Targaryen, who are the ethnic rulers of Valyria, had a premonition of a disaster called the 'Doom' that would befall their city, causing them to flee to Dragonstone. Both in the show and books, the ruins of Old Valyria are shown to be a post disaster site, and the book further emphasises that the Doom left Valyria completely uninhabitable:

Lakes boiled or turned to acid, mountains burst, fiery fountains spewed molten rock a thousand feet into the air, red clouds rained down dragonglass and the black blood of demons, and to the north the ground splintered and collapsed and fell in on itself and an angry sea came rushing in.

There are some GoT enthusiasts who dispute that it is a perfect allegory for climate change. Some have argued that the scene described above is also reminiscent of the time in which George R R Martin began to write the fantasy series. Back then, nuclear war was haunting the imaginations of people as a much more tangible threat, and there was little awareness of climate change. The vision of a apocalyptic Valyria is similar to our imagination of nuclear ravage, and to a certain extent similar to the bleak devastation left behind in Chernobyl, disturbingly similar to the description of the television show's Valyria that features a beautiful green maze but with diseased men looming.

Regardless of whether you agree if *Game of Thrones* is an allegory for climate change or nuclear war, one thing remains clear – its fantasy world is not so much fantasy as rooted in our present reality. Perhaps GoT is just a metaphor for our fears of self-destruction. I would agree the latter more strongly, considering how the series ending self-destructed itself with too much in the way of 'novice fantasy' in my opinion. A rushed last season, and a huge disservice to its female characters that had gradually emerged as the cornerstone of the show's appeal and success, but I will say no more to avoid any spoilers for those few among the global population who are yet to watch the show. I will say only this: prepare yourselves for seven magnificent seasons and a trash eighth one.

FEELINGS OF INDIA

Gazala Khan

When she was a young child, my mother, who is Belgian, developed a profound fascination with India and all things Indian. She recalls the exact moment, in second grade, when she took out a book from the library about the vast and populous country. Upon reading, she fell completely and utterly in love with all that she discovered about the nation, its people, history and traditions. This was no mere oriental fantasy, although she did dye her Barbie's hair black using ink from her cartridge pen and wrapped her in a make-shift handkerchief sari. So began a life-long love of India, inspiring her to study Indian culture at university in Belgium. To the surprise of no one, it was here that she met and married an Indian man and went on to have a few Belgian-Indian babies – my siblings and me. After graduating from university, my mom ran a little shop in Ghent selling Indian art and furniture that she imported. She named it Begum, a name that I adored because it was such a simple word yet rich with meaning. Begum, across South Asia, refers to a female of high ranking, a royal, or aristocrat. The name was a nod to the Begums of Bhopal, the four women who reigned over the Princely State of Bhopal, now part of Madhya Pradesh, where my father was born and raised. I always loved hearing stories of my mom's shop because, even though it was so hard to imagine my parents running a business, it sounded so romantic, magical even.

We eventually moved to the United States, and I must admit that in stark contrast to my mom, my interest in my Indian heritage was minimal. I suspect now that this was due to a misunderstanding on my part about how the world works. My theory was, I *am* Indian, therefore I don't need to learn, read, or hear about it because I should intrinsically know the politics, history, and culture, without seeking it out, because it is in my blood, and runs through my veins. I had the feeling of India, the smells, colours and sounds, already in me, having spent months of my childhood visiting family in Bhopal. This *feeling* I had for India, left me with a sense that I knew all that I needed to know, which ultimately meant I grew up not actually having learned anything about this aspect of my lineage aside from a very vivid image of what Shah Rukh Khan's abs look like.

To this day, most of the facts that I have absorbed about India come from a course I took in college called The Study of India. Don't get me wrong, I wanted to feel connected to my culture. I tried. I sat through endless Bollywood movies in an attempt to find something that would connect me to my Indian cousins and friends, but I found them long, boring and silly. I asked my parents to teach me to cook Indian food. 'Maaaybe, this much daal. Two fingers of water above the daal'. I watched my dad cup his hand and pour salt into it. Eyeballing everything, sometimes forgetting to tell me an important step, it took me years to make decent daal. Nothing seemed easy in my attempt to embrace my Indian identity and after a while the effort seemed insurmountable. I envied my younger sister, who loved Indian films, began Indian classical dance at age three and stuck with it into adulthood. I was much too shy to dance, but more than that, I had another more pressing concern - to assimilate. I wasn't born in the States like she was. I was a foreigner, and even though I pass for white, when I was a kid I had a Belgian accent, and I was probably the only kid in my elementary school who was part Indian and being raised Muslim. So there was a push and a pull, a confusing sense of pride mixed with shame. American kids told me to go back to where I came from, called me an immigrant as an insult. I don't think I consciously chose to reject my Indian side, but these things must have had an effect. Despite all this, I was happy to hang up saris as curtains in my room, and decorated my room 'Indian style' with my mom's antique Indian chairs and mirrored pillows. I loved the aesthetic. Perhaps this was a mirror to my relationship with the Indian part of my sense of self – objectified and a little detached.

A few years ago I picked up a book from my mother's extensive library, once again reaching for some knowledge about the culture of my father and that side of my family. It was a pretty book detailing the Mughal era through the lens of their patronage of the arts, in particular the exquisite miniature paintings produced during that period. I read it slowly and carefully, trying to understand the significance of these rulers. Having heard my mother refer to them with great reverence, and knowing the Begums had such an impact on her that she named her shop, her livelihood, her attempt at sharing her passion with the world, after this word, these women, I desperately wanted to understand who they were and to be impacted by the history too. Having always loved art history, something clicked and the book piqued my interest in the Mughals - I now had a rudimentary understanding of who these rulers were and their importance in world affairs.

Sikeena Karmali, *The Mulberry Courtesan*, Aleph Book Company, New Delhi, 2018

So it was after establishing this sense of connectedness that I had been seeking for so long that I picked up Sikeena Karmali's *The Mulberry Courtesan*. I was excited to be delving further into the world of the Mughals, of gaining a little more insight into the history of my family on my father's side, and devoured the story with enthusiasm. It follows the life of Laale, a young Afghan woman descended from a noble family who ends up serving as a courtesan to the last Mughal emperor of India, Bahadur Shah Zafar. I anticipated an epic tale of historical fiction with a strong and empowered female protagonist, all set against the stunning backdrop of nineteenth century India. The book does not disappoint when measured in value as a work of historical fiction. It portrays the elegance of a fading resplendent era through breathtaking descriptions of court, ritual and attire. However, my desire for a story of female emancipation in the context of a patriarchal society was far from realised. Perhaps it was an assumption made on my part from reading the blurb, which promised a 'fiery, independent, and beautiful' main character so it could be the case that my expectations were too high. But, whichever way I approached the text, my reactions varied from confusion to being rather appalled.

The story begins during Ramadan, and Laale, an exceptionally attractive, cultured and pious Muslim woman, is fasting. She is napping on her front porch, and her grandmother comments, as if by way of premonition: 'Look at her, lying there shamelessly with her *shalwar* hitched up past her shins, her dress tight around her breasts that are spilling out from the neckline, and her face glittering in the sun. Some hungry man is going to come and devour her!' Which is exactly what happens. I couldn't help but cringe at the way this blatant victim shaming played out, although I reminded myself of the concept of honour that has so inextricably been bound to women's bodies since time immemorial. Yet, although I am aware that this story is set in the past, and that Karmali is relating to the reader how scandalised the grandmother was by Laale displaying her body in public, I found the implicit link between beauty and licentiousness, uncomfortable, even offensive. The stories we are told through artistic mediums, whether novels, films, or myths, shape our perceptions and inform our beliefs. If one is telling a story in our contemporary times, even a story set in the past, the writer must bear some responsibility to attempt not to reinforce regressive ideas about what women should expect if they happen to be born with a beautiful body and are not ashamed or chaste enough to hide it. This idea, that Laale is just too beautiful for any man to resist is repeated multiple times throughout the book, and perpetuated the first time she is raped. This scene reads like an erotic fantasy for bored housewives. Laale isn't conscious, she is passive and unaccountable, bearing no responsibility in the event, aside from her alluring smell, and the 'mounds of her breasts tugging at the bodice of her *kameez*,' and her slack and parted lips that so overwhelm her captor with arousal that he rapes her. The idea that men rape women because they just can't help themselves demonises the vast majority of men who would never commit such an act and feeds into the narrative of the hypersexualised yet repressed Indian male who is unable to control his urges at the slightest glimpse of an unveiled woman.

Despite my growing unease, I was compelled to keep reading, because I was by now invested in the fate of Laale. Karmali heightens tension throughout the novel and uses suspense to keep up the pace of the story as we follow the plight of our heroine. It is through a series of events caused by her extreme beauty that Laale ends up as a royal courtesan. With her extraordinary intrinsic powers of poetry and dance, she wins the Emperor's heart and is spared the life of an ordinary courtesan, which means she will not be forced to have sexual relations with anyone against her wishes. There is much emphasis placed on Laale's sexual integrity from many different angles, which only serves to reinforce the misogynistic ideal of female purity. While on the face of it the idea of a woman using her attributes to reach the highest echelons of court life could be interpreted in a positive way, the manner in which it happens in this story is deeply depressing. Laale has no choice in the body she possesses. She did not work hard to become a great poetess, she blurted out a few lines whilst in the throes of a passionate dance that all came completely naturally to her, and they happened to be melodious enough to impress the royal poet and king. She has no agency and certainly did not choose to become a poet or dancer. She did what she had to do in rather awful circumstances and by some quirk of fate excelled, but more along the lines of a superhero with accidental powers than a strong female character. In fact it would be more accurate to describe her as a tragic figure.

There is one moment in the book where Laale reflects on the fact that like her, all the other courtesans must have had a history, a family, and wonders what journeys they undertook that brought them to this place. It is a rare moment of introspection for Laale, and one that does not last quite long enough to lead her to any real conclusion. At this point in the book the reader is almost encouraged to envy Laale. Residing in a magnificent palace, eating the finest food, and receiving dance and poetry lessons that she doesn't actually need because she is naturally so talented. She is living the dream, a woman that others will aspire to be like. But the dream does not last, because the Mughal empire is on the verge of downfall and her life of luxury will come to an end. Karmali illustrates the cruelty and deceit of the British well, leading to a palpable sense of frustration at the injustices that lead to the crumbling of the Mughal era. I wish Karmali had been able to create a similar effect in relation to Laale's predicament and the treatment of females, but her attempts fall flat. Perhaps it was out of a desire to make Laale seem strong and independent that she is given very little spoken dialogue and almost no inner dialogue. This results in a sense of detachment and despite the fact that Laale endures horrible things and manages to accomplish some extraordinary feats, I never felt much sympathy for her. I did, however, gain an interest in the events that lead to Shah Bahadur Zafar's demise and the East India Company's ascent to power.

My mother is finally living out her dream of living in India, dividing her time between Bhopal and Michigan. Recently, I had a conversation with her about the Mulberry Courtesan. She admitted that historical fiction was not her cup of tea, and recommended that I read some non-fiction works about the Mughals by William Dalrymple if I wanted to enrich my knowledge of Indian history. After describing the plot of *The Mulberry Courtesan*, my mom said it reminded her of the book *Umrao Jaan Ada*, and recommended I read it and watch the movies that are based on it. She was trying to recall who was in the film, maybe an actress called Rekha played the lead and I said 'Mom, the only Indian actor I know is Shah Rukh Khan' to which she jokingly replied 'What! What kind of Indian are you?' Good question. I'm working on it.

BETWEEN TWO WIVES

Samia Rahman

Polygamy is a contentious and often sensationalised topic yet audiences around the world seem to have an unlimited appetite for it. Prurient interest in the domestic arrangements of a man with more than one wife does not seem to have been dampened by the recent Western penchant for polyamorous relationships. Social media influencers and lifestyle blogs are shouting from the rooftops about the death of monogamy and the positivity of embracing multiple partners as life expectancy increases. I can barely cope with one significant other, never mind another, so can't quite see the appeal. Don't get me wrong, this is still very much a niche arrangement. What I find curious though, is the way it is treated in different contexts. There is an assumption that polygamous marriages in the Muslim world are exploitative of women. Yet the narrative surrounding polyamoury practised in other societies is couched in the rhetoric of sexual liberation and freedom from repressive convention. The truth, as is often the case, is complex. It is a shame that Hussain Ali Lootah's *Between Two Wives* does little to deconstruct the clichés and stereotypes in which the issue is entrenched.

As I picked up the novel I promised myself I would keep an open mind. I had an inkling this would be difficult but challenged myself to defy all cynicism. Even so, I could not have guessed what an impossible task this would prove. Suppressing every instinct to eye-roll, I scanned the spiel on the jacket. Apparently, this is the author's semi-autobiographical account of his 'turbulent inner struggles' as a man with two wives. Semi-autobiographical, I'm assuming, with all names changed to protect the innocent. And there certainly are many innocents in this tangled web of deceit, self-deception and misogyny. Our dear narrator, I am afraid, is not one of them.

It could be a plot-line from a Pakistani drama or a Netflix Original Series set in Egypt. Married man secretly takes second wife and finds the fantasy doesn't live up to the reality. At all. In fact he discovers it is terribly hard work maintaining two families and keeping them hidden from each other. His actions deeply wound those he professes to love and shatters the worlds of women who trusted him implicitly. But this is not a story about the aggrieved. The women and children who have been led a merry dance by the husband they believed in, are merely bit characters to the star of this performance. Because this book is all about him, his feelings, his anguish, his protestations and justifications. His is the only voice, we are led to believe, that matters. If this is meant to be a serious 'insight into the often misunderstood notion of polygamy' then I despair. Because despite the author's attempts to make himself vulnerable in this narration of his life's events, he inadvertently reinforces patriarchal stereotypes and throws a spotlight onto the cruelty and injustice of the choices he has made.

Lootah masquerades under his embellished pseudonym Yusuf, our protagonist. He describes how, as a young man, still a student of law at university in Dubai, he makes the decision to get married. His family are shocked but try to be supportive. He has chosen his bride, too. We are given no indication of how he and his first wife, Aliyaa, met and why he was so intent on marrying her. Was it the innocent young love of idealistic youth? Or a match arranged by a respected authority or spiritual advisor. We never find out. There is no evidence of any depth to their relationship. All we are offered is biographical detail of an orphan girl and her faltering start in life. We are even, tellingly, told that upon marriage, 'Aliyaa's story was about to begin'. The decision to marry was one that Yusuf made and insisted upon. Yet, his inability to take responsibility for his actions leads to barely-repressed reproaches to his parents. He asks why they didn't stop him from entering into such a serious commitment at his tender age. He cursorily blames them for choices he and he alone has made:

'They did not argue with him despite the seriousness of his decision. They did nothing to try and talk him out of his resolution. They did not even offer him any advice. He was only twenty-three years old... They approved of his unreasonable decision in an attempt to justify his behaviour. Even though Yusuf was resolute, he wondered why his parents were that lenient. Why didn't they speak up? Why didn't they say a word about his timing, about his choice? Why didn't they try to lay other options before him?'

Hussain Ali Lootah, *Between Two Wives*, Motivate Publishing, London, 2016

Yusuf perceives himself as a pawn of destiny and fate, as if he is merely a passenger in the story of his life. His emotional fragility and sensitive temperament as an adolescent are alluded to. He has 'hysterical fits', something he believes to be caused, or at least exacerbated, by the relentless teasing of his brothers, and is prescribed Valium to calm him down. He is always a victim of life's orchestrations, endlessly oppressed by the actions of others. The stress of being a husband to two women even takes a toll on his physical health. There is no mention of the impact it must have had on his wives and offspring.

The young couple's life together has begun and we learn a great deal about how Yusuf fares as a husband but little about Aaliya and whether or not she thrives in her role as a wife. The presence of a jinn in their home and in their vicinity is a recurring theme in the novel and I cannot help but wonder whether this uncritical acceptance of the phenomenon belies the truth of Yusuf's mental wellbeing. Again, all responsibility is advocated to a third party, this time a supernatural malevolent force.

Life is progressing until the world they have built for themselves is thrown off kilter by a chance encounter, not that Aaliya is aware, of course. The narrator's description of Yusuf's chance meeting of his second wife, Maria, unfolds in Lootah's typically figurative manner, alluding to this and inferring that but with scant details. The scene is tantalising, and whereas elsewhere in the book the allegorical style of writing is frustrating and evasive, the sequence of events that lead to that fateful moment is heavy with symbolism. The name of the country is not identified, only that it neighbours Ecuador, Maria's place of birth. The two groups that meet are not detailed in any way. We are given nothing with which we can paint a picture in our minds. Yusuf and Maria are affected deeply by each other, their lives have irrevocably changed. But we have no insight into their conversations, the flowering of their love. All we know is that Yusuf's torment at living two lives has begun.

While I no doubt feel great sympathy for Aaliya and the deceit she endures, my pain for Maria is equal. She leaves her home and follows the man she loves across the world to an unfamiliar life and culture. She does everything she can to conform to his expectations only to find that his protestations of devotion have been a sham. Yet she forgives him and tries to accommodate his selfish behaviour. She leads a solitary, miserable life, confined to the four walls of the basic accommodation he has secured for her, never questioning why his visits are intermittent or why she has never met his family or friends. All we hear are repeated laments of the emotional and financial burden he faces in maintaining two households. He is not a wealthy man and Maria is impoverished, sometimes left with little to eat. The hardships inflicted upon her are heart-breaking, and her acceptance of her situation incomprehensible. We learn of Yusuf's feelings of guilt and hear her occasional complaints, but she has no voice, no platform upon which she can be heard. Her thoughts and feelings and opinions are, like Aaliya's, imagined, assumed and delivered to us via the medium that is our narrator, Lootah, the man who is the source of the injustice.

The erosion of a woman's voice, the negation of her essence, her being, her existence; rendering her significance only in relation to the theatre of a man's life. This is the message of this story of denial and self-delusion. The narrator is at pains to point out that the taking of more than one wife hardly raises an eyebrow in some cultures. But this is a dying custom, met with increasing disapproval and a greater acknowledgement of the inadequacies of the contract. Upon revealing to Aaliya that he has secretly married another he undermines her outrage and that of her family by quoting his father's supportive counsel 'that such a tradition was an integral part of his heritage... Why then make a fuss out of Yusuf's issue? Why all this trouble? Why couldn't he live the way he wanted to?' Such insensitivity to the feelings of the women and children he has betrayed proved infuriating to read.

Yusuf's self-absorption does make way for occasional glimpses of genuine insight. An interesting dialogue between himself and a European business associate is one example. The two men debate the hypocrisy of demonising polygamy while many men in the West have a string of mistresses. Although I am troubled by the practise of taking more than one wife I would never go so far as to deny the patriarchal structures that blight the lives of women in the West. No nation, religion, culture or community has the monopoly on misogyny.

There is little else in this book that I found illuminating, although certainly, it sheds a spotlight on a fascinating moral conundrum, and the author's searing honesty is raw and uninhibited. We are privy to the internal thoughts and motivations of a traditional Emirati man and the subject of polygamy is not approached salaciously. If anything the book is devoid of the frivolity of sensation. It is preoccupied only with regret, recriminations and victimhood. At least Yusuf concedes he is not the only victim, with some caveats. 'I apologise. I apologise to those who fell victim to my life: myself first.' He and his families are not victims of his actions, they are victims of destiny. Because of course, as anyone reading this self-absorbed fable will know, he is surely absolved of all blame.

ET CETERA

LAST WORD

ON OUR FUTURE

Mohamed Aidid (age 15) and *Fareha Rahman* (age 15)

This is an emergency. Time is running out and everything you know, everyone you love is at risk. Your friends, your family, the entire global family. Because we are a family, we are humankind: interconnected, culpable and with the power to affect change. Not just the power but also the responsibility. Without immediate and decisive action we face extinction. We know that some people find this difficult to grasp, or even believe, but really, this is our situation. It could not be clearer or more bleak. Climate breakdown and ecological collapse are a direct and impending threat to our civilisation. This is not an exaggeration. We are deadly serious, and we are determined that our generation will not accept the inaction of those who have our future in their hands.

We are already in the sixth mass extinction of life on Earth. Did you know that 200 species are lost to our planet every day? Lost forever. Just think of all those creatures you have lived alongside that we will never be able to lay our eyes on. Every month we are experiencing more and more extreme weather and this is wreaking havoc in countries everywhere. Floods, wildfires, droughts, and crop failures are ravaging communities that have never previously experienced anything like this before and are having their livelihoods and their lives terribly disrupted. Desperate conditions are leading to food shortages and this is causing social upheaval and conflict. The consequences are just so dire with hundreds of thousands of lives lost to climate breakdown every year.

There are so many issues – structural inequalities and injustices that mean those in power choose to ignore the seriousness of the threat of climate change because it is not in their interests to do so or because it limits their profits. But what young people like us are calling for is not just to change this or that, anything, but to change *everything.* We have to change this system based on industries and technologies, greed and naked exploitation, that are intrinsically destructive to the environment. We need to move forward in support of the most basic of universal values – the right to a future, the right to safeguard the planet we will inherit. We have to act because we have no choice. We do not have anywhere else to go. There is no Planet B. The earth is the only home we will know and we must do all we can to save it from the neglect of those who have gone before us. We are angry because governments have failed to act accordingly, continuously prioritising profit over life. We simply cannot understand why they are not doing more to stop climate collapse.

The warning was sounded as far back as thirty years ago when the United Nations joined forces with the world's scientists to clearly state that carbon emissions needed to be drastically cut if climate collapse was to be averted. It is unbelievable to us that three decades later emissions have not been cut at all but instead have risen by 60% without any slowing down of this rise. Experts are concerned that there will be more tension in the Middle East as a consequence of climate collapse. A recent report in the *Guardian* described how droughts in Guatemala due to lack of rainfall were forcing citizens of that country to make perilous journeys to the US to escape poverty and starvation. Upon arrival they are humiliated and treated like criminals. This breakdown of humanity is only going to get worse if we don't take collective responsibility and pursue radical solutions to combat environmental disaster. The most frustrating aspect of all this is that change to hopefully stop the worst of the disaster could still be possible if there is the political will.

We know we couldn't just stand by and do nothing. One of us – Mohamed Aidid – is the Youth Mayor of Bristol; the other – Fareha Rahman – is a volunteer for Islamic Relief. We realised that it is our responsibility to lead by example and play an active role in the youth strike for climate. We handed in a petition to the mayor and his office from young people all across Bristol and made a short speech about the undeniable danger of climate change and how we must be fighting for climate justice. We pointed out that Anthropogenic Climate Change is caused mostly by human activity. Such climate change has a massively detrimental impact on human society and on all natural systems. We suggested that climate change is blighting the lives of people in the developing countries, displacing them and turning them into climate refugees. It may lead to the ruin of their economies, exacerbate health problems that already pose a major burden on vulnerable communities, and will affect food security. Hurricanes, floods and typhoons are becoming more frequent and more severe, the polar icecaps are melting, and temperatures in certain parts of the worlds have now reached a level way beyond the human capacity to bear. We must understand we currently live in a world in which overpopulation is an issue and there is an inaptitude to adequately provide for those vulnerable and less fortunate. We currently have the chance to prevent things from getting worse but we aren't doing anything about it. The cost of prevention is nothing in comparison to the cost of fixing the damage. Some forms of damage, like a life, are irredeemable. Another frightening consequence is that millions of young people will never get to fulfil their goals and dreams – thanks to the mistakes made by past generations and their apathy and unwillingness to deal with the issue at hand. The implications lie ahead, and the need to agitate for action falls upon us.

We became aware of the YouthStrike4Climate and Extinction Rebellion movement from social media, after seeing promotions and preparations for the strike. Of course the demographic of the strike was us young people, and it really captured our imagination, with the strike becoming a big topic amongst our peers. After having contemplated whether we should attend the strike or not, we decided that making an active gesture by being involved and supporting the idea of a necessary change, was more important than the French lesson we would miss.

Our school was very supportive and understood why we felt the need to strike. The consequence of a day of absence from our classes, in our eyes, was nothing compared to a day of effort taken to save our future. Our school appreciated the fact that it is time us young people had our voices heard.

With such activism, we hope to achieve more awareness of climate change and the need for climate justice but also the need for more diversity and more accessibility within climate justice movements. Climate change is an issue that touches all races and cultures in some way, shape or form and it is an issue that everyone should be able to relate to. We can all make a difference. Even by doing small things like bringing awareness during school assemblies or by organising petitions. Collectively and united we can make a big difference. For change to occur on such a momentous issue, collective action is essential – and that is what we aim for.

Social media has played a massive part in energising and encouraging people to join the movement. Although social media has some horrible negatives but one of its assets is its ability bring people together and that's why so many people are enthused and riled up about climate collapse.

Muslim communities can do more to engage with the issue of climate collapse. Muslim countries are amongst the first to be affected by the severity of climate change – many have begun to feel the consequences of temperature rises, water shortages, desertification, sea level rises and massive flooding. Poorer Muslim countries do not have resources to cope with the terrible ways in which climate change will disrupt the lives of their citizens. So it is important for us Muslim not just to be fully aware of the consequences of climate change but also to take action to break and reverse the trends.

What are YouthStrike4Climate and Extinction Rebellion calling for? We are in a time of an intense crisis. Sheer desperation is forcing us young people to act, yet we feel ignored by those in power. We young people demand that governments take immediate action. And we will continue to protest and rebel and to do what we need to do to fight for our future.

There are numerous brilliant, detailed proposals for how things could change, how problems could be solved. But we can't make changes without taking our power back, getting more connected to each other and to the wider world. We reckon wholesale system change needs a mass movement of people willing to take disruptive, loving and effective direct actions, saying no to the destruction and deadly policies, and yes to life sustaining alternatives. An inclusive uprising will have many ways for people to get involved. It must be effective and sustainable, it must be fun.

Our demand is that the Government must tell the truth about the severity of climate change, and just how deadly our situation is. It must reverse all policies not in alignment with that position and must work alongside the media to communicate the urgency for change including what individuals, communities and businesses need to do. The Government must enact legally-binding policies to reduce carbon emissions in the UK to net zero by 2025 and take further action to remove the excess of atmospheric greenhouse gases. It must cooperate internationally so that the global economy runs on no more than half a planet's worth of resources per year. We demand a Citizens' Assembly to oversee the changes, as we rise from the wreckage, creating a democracy fit for purpose.

Our International Non-Violent Rebellion Against the World's Governments for Criminal Inaction on the Ecological Crisis has begun.

OBITUARY

'UNCLE IDRIS'

M Iqbal Asaria

I was invited to the inaugural meeting of the Third World Network in 1982. When I arrived at Kuala Lumpur airport, the Chinese immigration officer looked at my British passport and then looked puzzlingly at me. After a pause, he told me that he was obliged to send me back to the UK as my passport had expired. This was before the arrival of digital passports and it was easy to make this kind of mistake. The immigration officer then asked me what was the purpose of my visit to Malaysia. I replied that I had been invited by S M Idris to a meeting in Penang. 'Visiting Uncle Idris, *lah*!' he said. He could not send me back, he declared, as Uncle Idris would be offended. He then found a rarely used provision in the law and gave me a temporary entry pass for three days and advised me to get my passport renewed at the British Consulate in Penang. When I visited the Consulate, the consular officer, a Chinese lady, upon hearing the purpose of my visit, rushed to organise a new passport. 'I respect Uncle Idris's work very much, *lah*', she said.

S M Mohamed Idris, who died in Penang on 17 May 2019, age 92, was a much respected and adored pillar of Malaysian society. Affectionately and universally known as 'Uncle idris', he had acquired an international reputation for his work and advocacy on citizens' rights and environmental conservation. Indeed, such was his stature that the city of Penang is now considering naming a road and a park after him.

Uncle Idris was born in Tamil Nadu, India, where he was educated at an Islamic madrasah. When his family moved to Penang, he attended a local Christian missionary school, and worked with his father, SM Mohamed Yusoff Rawther, who had started a shipping company in 1915. His father died during the Japanese occupation of Malaya (December 1941-February 1942), and Uncle Idris took over the shipping business and a small jewellery shop. Almost immediately he started a union for shipping workers' rights: Penang Lighters' Association. It marked the beginning of a life-time of campaigns, advocacy, and struggle for the rights of workers, the homeless, squatters, consumers, and flora and fauna.

At his funeral, Anwar Ibrahim, President of the People's' Justice Party and the leader of the Pakatan Harapan coalition, described him as 'a simple man, something that we rarely see these days, and being a pious Muslim, his approach was imbued with patience and simplicity'. Indeed, Uncle Idris radiated piety and simplicity. He was a proud Muslim who was always dressed in an immaculate white sarong and tunic commonly adorned by South Indian Muslims, accompanied by a *Songok* (a black velvet pointed hat) or a white cloth cap. This signature attire always made him instantly recognisable. He was also a man of meticulous discipline. Even at the age of 90, you would find him at his desk at 8.30 sharp in the morning reading the papers. Computers and smart phones were not to his taste.

But Uncle Idris' simplicity clothed a profound intellect. He was an extremely well read person and a sharp critic of dominant paradigms. His favourite topics for demolition were the exploitative nature of Western civilisation, the oppressive structures of capitalism and global finance, and the prevailing ways of knowing and doing. An hour in the presence of Uncle Idris was like attending an intensive, and demanding, postgraduate seminar. So Uncle Idris did not just work for the rights of the downtrodden and the marginalised, he also worked tirelessly to dethrone the accepted consensus on knowledge production, the paradigm of development, the Eurocentrism of disciplines and universities, and the unjust global economic system. He established the Consumer Association of Penang to champion the rights of the underprivileged; Third World Network to bring thinkers, scholars and activists from the global South together to fight the dominance of the West; and Sahabat Alam Malaysia, to campaign for the conservation of the environment and ecologically sound policies.

He was a true polyglot: he spoke a number of intellectual languages, not least profound simplicity, and understood a string of complex issues! A flavour of his acute intellect was provided at a meeting in February 2002, organised to launch another Uncle Idris project, Multiversity – a network of scholars, academics, schools, colleges and universities dedicated to stimulating 'a fresh look at learning for the self, for the community and for the environment'. Uncle Idris raised a series of questions about the Western knowledge system: 'Whose knowledge system is this? What is its purpose? If we did not create it how can we be asked to claim it as our own? Why do we reject our own (traditional) knowledge, valid and result of centuries of experience, for such knowledge? What are its assumptions? Have these been scrutinised by any of our intellectuals – secular or religious? What about our own knowledge? If it is valid in our own countries, why is it not valid elsewhere? Is our knowledge valid only if re-investigated or patented within the perspective of the modern system?'. He went on to suggest that 'the world system has perfected a method of training and selection that enables it to recruit for itself, apparent, the brightest and the best, and, after selection, to use such recruits against the interest of the rest'.

The Consumer Association of Penang (CAP) was established in 1970 to provide 'a voice for the little people'. Upfront, CAP fights for the right of every consumer to food, housing, health care, sanitation facilities, public transport, education and a clean environment. But behind the scenes, CAP carries out research, educational and representational activities to influence policy makers to give priority to basic needs. It handles about 3,000 to 4,000 complaints from the public every year on issues ranging from poor quality products and services to food adulteration and housing. CAP's campaigning news magazine, *Utusan Konsumer* has four bimonthly editions each published in English, Malay, Chinese and Tamil. CAP championed environmental causes from its inception. After the infamous incident at Penang's *Batu Firringi* (Foreigners Beach), which became highly polluted by toxic dumping from the then rapidly developing semi-conductor industries, CAP and Uncle Idris mobilised public opinion for the creation of a Ministry of the Environment. It was duly established in 1975.

During the 1970s and 1980s, CAP was renowned for its seminars and annual conferences, attended by scholars, scientists, journalists, activists and intellectuals of all shades of opinion and background. It was, in fact, a regular meeting place for the radicals and dissenters of the world. Whatever the theme and subject of the conference – which ranged from development, science, technology to traditional knowledge systems, globalisation and the media – the emphasis was always on what the Third World itself could and should do to improve its situation, how developing countries could free themselves from underdevelopment and western domination. Uncle Idris would preside over the conferences but allow free reign for the participants to change the programmes as they wished. His main concern was that the conference should be much more than simply occasions for debate and discussion: the participants were required to produce teaching material for universities, and policy recommendations which could be used as spring board to launch conscious raising or reformist campaigns directed towards communities, industries and local and national government. A particular incident at the 1986 conference on 'Science in Crisis' is noteworthy. Ziauddin Sardar was asked to speak on 'What is Wrong with Science?'. He asked the Chairman to leave the podium, rearranged the furniture, and invited Indian historian of science, Claude Alvers, to join him. There followed a riveting two-hour wide-ranging conversation on science and exploitation, science and racism, science and sexism, science and militarisation, the inability of science to promote equality, the linkage between science and economics, science education, and appropriate technology. The conversation was recorded and became the basis for the 'Penang Declaration on Science and Technology'. Later, Sardar turned the proceedings of the conference into a highly cited book: *The Revenge of Athena: Science, Exploitation and the Third World*. The dedication lovingly states: 'For Uncle Idris'.

CAP was also instrumental in setting up the Third World Network (TWN) which brought together leading activists and thinkers from the global South. Martin Khor Kok Peng, economist and close colleague of Uncle Idris, took charge of the organisation, which evolved into one of the leading institutions of the Global South. TWN's annual meetings, held at the amicable setting of the Lone Pine Hotel on Penang's *Batu Firringi*, attracted leading thinkers from the developing world. Indian philosopher and cultural critic Ashis Nandy, Pakistani critical theorist Iqbal Ahmad, Bangladeshi founder of Grameen bank, Mohammed Yunus, and Moroccan feminist Fatima Mernissi were regular participants – along with noted intellectuals and thinkers from the Philippines, Thailand, and South America.

I was lucky to be present from the first meeting in Penang and attended most of them over the next ten years. We debated the activities of global agribusiness and their nefarious plans to mass market toxic chemicals, terminator seeds and the like. The recent award against Monsanto for the harm caused by its *Roundup* herbicide was already in our sights in those early days, as was the damage caused by unsustainable logging and deforestation. So was the movement to push monoculture varieties of key crops like rice and wheat, exposing the producing countries to widespread disease and depletion of rich diversity needed for resilience against particular pests.

We realised that these policies were pushed by a number of key institutions and individuals who were supportive of the agenda of multinationals corporations. So we decided to take the battle to its source – the World Bank and other UN organisations. At the World Bank, TWN became an active member of the World Bank NGO committee. As chair of the WB-NGO Committee on behalf of TWN, I was part of the effort to question the Washington Consensus promoted via the Bank's Structural Adjustment Programmes to extract resources from the global South without regard to any safeguards. The world saw the results of these policies in such incidents as the December 1984 Bhopal disaster when a Union Carbide facility sent toxic gas discharge resulting in thousands of fatalities and injuries. The US multinational was hardly held accountable for this perfidy. India was prevailed upon by Washington to accept paltry compensation and not make a fuss lest it loses US support at key international fora. As a result, The WB-NGO interactions became more forceful. Although not the sole work of the WB-NGO Committee, the pressure mounted on the World Bank and UN institutions to take poverty reduction seriously and eventually led to the advent of the UN Millennium Development Goals (MDGs). These have now been extended into the Sustainable Development Goals (SDGs) which are the key focus for any contemporary discussion on sustainability of the planet.

As our capacities built up, Uncle Idris's relentless push led us to explore wider vistas. The TWN got involved in the General Agreement on Trade and Tariff (GATT) negotiations in the Uruguay Round, spanning from 1986 to 1994. Here again the battle was arduous. A typical case in point was the move by tobacco companies to promote smoking in developing countries to compensate for their stagnant markets in the developed world as awareness of the harm caused by smoking was gaining ground. The companies had identified young females of child bearing age as the most fertile market for their wares. When the local NGOs in Thailand objected to these vile policies, the companies threatened to sue under GATT rules and penalise Thailand with other tariffs. This was an eye opener for the TWN campaigners and showed that the scale of the problem was massive and enormous resources were needed to combat it. The TWN continued to influence key thinkers and decision makers in the global South to champion these causes.

Yet more was to come. The TWN had organised a continent wide tour of India with its key experts to present its work and policies to a wide range of academics and civil society. We toured Delhi, Bombay, Madras and Calcutta. The support and resonance for our ideas was tremendous. The group was invited for dinner at an upmarket restaurant in Calcutta. Uncle Idris was dressed in his usual dignified South Indian attire. The doorman at the restaurant stopped him and authoritatively said, '*Saab Lungi* not allowed'. He was effectively saying that the restaurant's dress code did not entertain his traditional, local attire. I have never seen Uncle Idris more furious and upset at this insult. For several days after this incident local officials had to suffer the wrath of his anger and disappointment at the prevalence of colonial attitudes in independent India.

Once the dust had settled, the incident led Uncle Idris to his final mission. This was the Multiversity project. Uncle Idris felt that universities and centres of learning in the South had to be decolonised. The colonised minds of our scholars and thinkers, argued Uncle Idris, were perpetuating colonial attitudes, policies and status quo. Tradition was, consciously and unconsciously, being downgraded and considered inferior. Radical new insights were not transmitted to new generations or society at large. A particularly noteworthy outcome of Multiveristy is the series of 'Dissenting Knowledge' pamphlets, edited by the Indian historian and cultural critic, Vinay Lal. I can see Uncle Idris' face light up at the titles of some of the tracts published over the last decade: *The Tyranny of Economics* (Roby Rajan), *Is Science Western in Origin?* (C K Raju), *Exorcising Anthropology's Demons* (Frederique Apffel-Marglin and Margaret Bruchac), *Digital Diploma Mills* (David F Noble), *White Studies* (Ward Churchill), *Necessary Terrorists:Why the West Hates Muslims* (Yusuf J Progler).

Uncle Idris, a self-taught man of elegant simplicity, taught, inspired and motivated so much to so many all over the world. Personally, I owe a tremendous debt of gratitude to Uncle Idris since our first meeting in 1982. Ever since that fortuitous encounter, he has been a true friend, mentor and inspiration who opened new vistas and changed my thinking. His constant fatherly admonitions, whenever I showed signs of slackening, were typical of his unique style of getting people to achieve their maximum potential. He would not take no for an answer. Visiting him would always bring forth a long list of tasks which needed to be done urgently to improve the condition of the marginalised.

We lost a proud and confident Muslim committed to serve humanity to the best of his ability. He inspired a whole generation to rise up to challenge and change the world for the better.

His loss will be felt not just in Malaysia but throughout the global South.

S M Idris, visionary, environmental campaigner, consumer activist, profound critic of Western Civilisation and Western modes of knowledge production, founder of Consumer Association of Penang, Third World Network, the environmental pressure group Sahabat Alam Malaysia, and Multiversity, and legendary peoples' champion, born 6 December 1926, died 17 May 2019.

SEVEN CLIMATE-DENYING WONDERS OF THE WORLD

Contrarians have always existed. Sometimes celebrated, other times ridiculed. Subversives who defy dominant thinking to advocate an alternative truth often find their views eventually appropriated by the mainstream. It was only a few decades ago that those voicing concerns about the dangers of 'global warming' were regarded with scepticism and dismissed as tree-hugging hippies. Now, alarm at the peril of climate collapse is widespread across continents and communities. Government complacency is provoking demands for action by most people who consider climate change an irrefutable fact. But, much like the flat-earthers who refuse to accept overwhelming scientific evidence that the world is round, there are those who cling on to the belief that climate change is at best a distortion, at worst an untruth. Climate change deniers are a minority, but their disproportionate influence is a reflection of their positions of power. Here are the Top Seven hindrances in the fight against climate collapse.

1. The Australian Electorate

It was touted as Australia's climate change election. Having endured the hottest summer on record, with temperatures peaking in the high-forties, surely the Australian population would rise up to act in the face of a dire emergency. Wildfires triggered by tinderbox conditions were occurring in places that had never been vulnerable to raging fires before. The coal-dependent energy supply repeatedly buckled under intense pressure as frequent power cuts compounded the suffering of those confined by the extreme heat to their air-conditioned homes. Farmers were forced to write off crops. Freak floods destroyed some areas while unprecedented drought blighted others. The choice was clear. Vote for the Greens and

Labor to avert catastrophe or continue with the Liberals and their reticence to take action. The electorate chose the latter. From coal mining communities seeking to protect their jobs to urban dwellers worried about taxation and house prices, short term economic security trumped long-term policy to tackle climate change. A nation in denial.

2. Alternative for Germany (AfD)

The far-right Alternative for Germany (AfD) party has pushed climate change high up on its policy agenda. But not in the way you might think. Its strategy appears to be to perpetuate the fears of communities who see environmental concerns as an attack by 'elites' and 'experts' on their freedom and jobs. This message, typical of right-wing populist movements across the world, seems to resonate with German voters in industrial regions who feel their livelihoods under threat from eco-activists. Referring to a hysterical 'cult of Greta', when talking about the work of sixteen-year-old Swedish climate strike activist Greta Thunberg, AfD dismiss the climate change movement as a brainwashing replacement for religion, and warn against the renewable energy sector's deindustrialisation agenda. Fortunately, they did not achieve the success they had anticipated in recent European elections and instead the Greens surprised everyone with huge wins in Germany.

3. Indonesia

According to a YouGov poll, conducted as part of the Cambridge Globalism Project, the country with the greatest number of climate sceptics is Indonesia, which also happens to be one of the world's largest emitters of greenhouse gases. Beating both Saudi Arabia and the United States, this accolade is nothing to crow about, particularly as Indonesians are already grappling with the havoc wreaked by unpredictable weather patterns. Woefully unprepared for what lies ahead, activists blame a lack of education and suspicion of the highly politicised nature of environmental debates for the general apathy and antipathy. Interestingly, the fatalistic Indonesians are not in complete denial that climate change is occurring. What they dispute is that humans are the cause.

4. Saudi Arabia

The oil-producing Gulf state continues to play a crafty game of cards in its efforts to protect its vested interests, while at the same time presenting a PR-friendly front on tackling climate change. Despite predictions that temperature changes will have a particularly calamitous impact on the Middle East, the Saudis have repeatedly resisted calls to adopt emissions targets. At the UN's climate talks in December 2018, Saudi Arabia joined forces with other climate-sceptic and high-polluting nations including Russia and the US, to obstruct any meaningful progress. Worse still, in a recent Twitter rant, the nation's former lead negotiator for the 2016 Paris agreement, a United Nations Framework Convention on Climate Change dealing with greenhouse gas emissions mitigation, adaptation, and finance, branded the agreement a 'big conspiracy' orchestrated by the 'climate mafia'.

5. President of Brazil, Jair Bolsonaro

The rainforests of Brazil have been described as the lungs of the Earth. Swathes of oxygen-producing, carbon dioxide-absorbing vegetation has stymied the harm caused by greenhouse gases. A beacon of sustainability, Brazil has done much in recent decades to prevent deforestation, stamp out illegal logging and increase renewable energy production. However, the election of the far-right climate sceptic Jair Bolsonaro, and his courting of the powerful agribusiness lobby could jeopardise all the vital environmental progress that was achieved. Likened to a conservative dictator, Bolsonaro is accused of homophobia, sexism and racism, but his populist crusade against corruption and promise of economic stability is what rural communities seem to want to hear.

6. Russia

Russia's president Vladimir Putin could not entirely accurately be described as a climate denier. This is despite making frivolous remarks that global warming would save Russians money on fur coats and increase wheat harvests in Siberia. He has, rather, played a cunning game of sidestepping calls for a reduction in emissions by proposing a strategy whereby adaption to climate change, and an attempt to profit from changing conditions, is the way forward. He has also found a perfect distraction from his own shortcomings in the buffoonery of the President of the United States, talking of whom…

7. Donald Trump

Where to even begin? When the US president isn't grabbing women by the pussy or sending thoughts and prayers to school gun shooting victims in lieu of tightening firearm controls, or describing Mexicans as rapists, he is defying all belief in his ignorant remarks about climate change. The president of the country with the largest economy and self-styled 'leader of the free world' has blunderingly implied climate change is a foreign conspiracy, is unthinkingly anti-science and makes ludicrous comments based on fantasy, for example claiming the sound from wind turbines has been linked to cancer. Supported by a Senate that is deeply vested in coal, oil and gas interests, Trump has appointed climate change deniers to senior positions in government. History is likely to look back at Trump as a pivotal figure in the catastrophic failure to address the danger of climate change.

CITATIONS

Introduction: Endgame by Ehsan Masood

On box tree caterpillar, see RHS 'Where did the Box tree caterpillar come from and what to do about it': https://www.rhs.org.uk/advice/profile?pid=760&awc=2273_1560906475_1073dc8f55032fcb75b1155aad7b7469.

On mass extinction, see National Geographic:
https://www.nationalgeographic.com/science/prehistoric-world/mass-extinction/

The IPCC special report on global warming warns that irreversible climate change could be just 12 years away:
https://wwwipcc.ch/2018/10/08/summary-for-policymakers-of-ipcc-special-report -on-global-warming-of-1-5c-approved-by-governments/

The Islamic Declaration on Global Climate Change can be viewed at:
http://www.ifees.org.uk/declaration/islamic-climate-change
-syposium/
and Yale University Environmental Performance Index can be found at:
https://epi.envirocenter.yale.edu/

On Science and Empire, see Zaheer Baber, *Science of Empire* (State University of New York Press, Albany, New York, 1997) and Patricia Fara, *Sex, Botany and Empire: The story of Carl Linnaeus and Joseph Banks* (Icon Books, Cambrdige, 2003). On Al-Mamun, see Michael Cooperson, *Al-Mamun* (OneWorld, Oxford, 2005).

See also: Ehsan Masood, *The Great Invention: The Story of GDP and the Making and Unmaking of the Modern World* (Pegasus Books, Singapore, 2016); and

Macartan Humphreys, Jeffrey Sachs and Joseph Stiglitz, *Escaping the resource curse* (Columbia University Press, 2007).

The *Barakah* of Water by Medina Tenour Whiteman

For translations of classic texts, such as Khayr al-Din ibn Ilyas's *Kitab al-fallah*, see: filaha.org; for more on permaculture visit supernatural-permaculture.com. Permaculture Research Institute has a permaculture map of ... Yemen at:
https://permaculturenews.org/2012/01/04/yemen-on-the-permaculture-map/

La Loma Viva website is at: lalomaviva.com. You can subscribe to Zawiyah at thezawiyah.org.

Too Close to the Sun? by Giles Goddard

Lynn White's much cited paper, 'The Historical Roots of the Ecological Crisis' was published in *Science*, 10 March 1967. See also Elspeth Whitney, "Lynn White Jr.'s 'The Historical Roots of Our Ecologic Crisis' After 50 Years", *History Compass* 2015, Vol.13(8), 396-410. doi:10.1111/hic3.12254. The quote from the Bible: Genesis 1. 26, 27.

The Hans Jonas quotes are from *The Imperative of Responsibility* (University of Chicago Press, 1984), p.1 and Hava Tirosh-Samuelson and Christian Wiese, editors, *The Legacy of Hans Jonas: Judaism and the Phenomenon of Life* (Brill, Leiden, 2008) p. 135.

I.G.Simmons quote is from *Global Environmental History*, (Edinburgh University Press, 2008) p111.

Walt Whitman's 'Songs of Parting' is widely available on the web as is the Pope's 8 March 2019 speech at the Vatican conference on sustainable development. The pdf of 2015 Islamic Declaration on Climate Change can be downloaded from: http://www.ifees.org.uk/wp-content/uploads/2016/10/climate_declarationmMWB.pdf

Gaia 2 by Christopher Jones

On global weirding, See John Sweeney, 'Command-and-control: Alternative futures of geoengineering in an age of global weirding.' *Futures* 57: 1-13 2014; and 'Global Weirding' *Critical Muslim* 17: Extreme (Hurst, London, 2016).

On Gaia theory, see James Lovelock, *Gaia: A New Look at Life on Earth* (Oxford University Press, 1979); *The Ages of Gaia* (W. W. Norton, New York, 1988); and *A Rough Ride to the Future* (The Overlook Press, New York, 2015).

On postnormal times, see Ziauddin Sardar, editor, The Postnormal Times Reader (IIIT and Centre for Postnormal Policy and Futures Studies, London, 2017, 2019).

Authors cited include: J Barker, 'Learning about the Limits to Growth.' *APF Compass* Special Edition, Limits to Growth: 2-6 2019; Donna Haraway, *Staying With the Trouble, Making Kin in the Chthulucene* (Duke University Press, Durham, NC, 2017); Mark Lynas, *Six degrees. Our future on a hotter planet* (National Geographic Society, Washington D.C., 2008); David Wallace-Wells, *The Uninhabitable Earth* (Tim Duggan Books, New York, 2019).

Thomas Friedman's New York Times article, 'Global Weirding Is Here', can be retrieved from http://blog.canacad.ac.jp/wpmu/kiaora/files/2014/04/English-Paper-1-Text-C.pdf; and the report on the effect of global warming on glaciers, 'Glacier in Russian Arctic Goes From Moving 60 Feet a Year to 60 Feet a Day' weather.com (8 April 2019) can be retrieved from: https://weather.com/news/news/2019-04-08-russian-glacier-moving-much-faster).

See also: Christopher Jones, 'Gaia Bites Back'. Special Issue on Human Catastrophes, *Futures*. Vol. 41, No. 10. December 2009; 'Sustainable Future.' *Encyclopedia of Sustainability in Higher Education*. (8 November 2018). W. L. Filho, ed. DOI: https://doi.org/10.1007/978-3-319-63951-2_245-1 2018; and 'When Things Fall Apart: Global Weirding, Postnormal Times, and Complexity Limits.' (2019). In *Building Sustainability Through Environmental Education*, L Wilson & C. N. Stevenson, eds. IGI Global. 10.4018/978-1-5225-7727-0.ch007; Bruno Latour and T Lenton, 'Gaia 2.0.' *Science* 361(6407): 1066-1068 (2018).

In Love and Rage by James Brooks

Rebellion Day Facebook page is at: https://www.facebook.com/events/bridges-in-london/rebellion-day/1758991460816073/

Roger Hallam's *Guardian* article, 'Wake up, Britain. We've been betrayed over Heathrow' can be read at: https://www.theguardian.com/commentisfree/2018/jun/27/britain-betrayed-heathrow-humanity-survival

You can listen to 'Je Suis un Sauvage' by Alfred Panou and The Art Ensemble of Chicago, (1969) on YouTube.

Corder Catchpool, *Quaker Faith and Practice* (fifth edition) is published by Quaker Books (Morley, 2013).

Denial, Deceit, Bullshit by Gordon Blaine Steffey

Myles Allen *et al.* 'Framing and Context.' *Global Warming of 1.5°C. An IPCC Special Report on the impacts of global warming of 1.5°C above pre-industrial levels and related global greenhouse gas emission pathways, in the context of strengthening the global response to the threat of climate change, sustainable development, and efforts to eradicate poverty*, 2018, https://www.ipcc.ch/sr15/chapter/chapter-1-pdf/.

John Cook *et al.* 'Quantifying the consensus on anthropogenic global warming in the scientific literature.' *Environmental Research Letters* 8 (2), 2013.

John Cook *et al.* 'Consensus on consensus: a synthesis of consensus estimates on human-caused global warming.' *Environmental Research Letters* 11 (4), 2016.

Theodosius Dobzhansky. 'Nothing in Biology Makes Sense Except in the Light of Evolution.' *The American Biology Teacher* 35 (3), 1973.

J.K. Farrell, K. McConnell & R. Brulle. 'Evidence-based Strategies to Combat Science Misinformation.' *Nature Climate Change* 9 (3), 2019.

Stuart Firestein. 'Doubt is good for Science, but bad for PR.' *Wired*, 2012, https://www.wired.com/2012/07/firestein-science-doubt/.

Harry Frankfurt. *On Bullshit*. Princeton and Oxford: Princeton University Press, 2005.

Harry Frankfurt. 'Donald Trump is BS, Says Expert in BS.' *Time*, 2016, http://time.com/4321036/donald-trump-bs/

David Fridley. 'Nine Challenges of Alternative Energy.' In *The Post Carbon Reader: Managing the 21st Century's Sustainability Crises*. Eds. Richard Heinberg & Daniel Lerch. Healdsburg, CA: Watershed Media, 2010.

Douglas Futuyma. *Evolutionary Biology*, 2nd ed. Sunderland, MA: Sinauer Associates Inc., 1986.

Stephen Jay Gould. 'Evolution as Fact and Theory.' *Discover* 2 (5), 1981.

Abel Gustafson *et al. The Green New Deal has Strong Bipartisan Support*. Yale University and George Mason University. New Haven, CT: Yale Program on Climate Change Communication, 2018.

Alexa Jay *et al. Impacts, risks, and adaptation in the United States: Fourth National Climate Assessment, Volume II*. U.S. Global Change Research Program, Washington, DC, 2018, doi: 10.7930/NCA4.2018.CH1.

Dan Kahan. 'What Is the science of science communication?' *Journal of Science Communication* 14 (03), 2015.

Edward Larson. *Summer for the Gods: The Scopes Trial and America's Continuing Debate over Science and Religion*. New York, NY: Basic Books, 2006.

Antony Leiserowitz *et al. Climate change in the American mind: December 2018*. Yale University and George Mason University. New Haven, CT: Yale Program on Climate Change Communication, 2018.

Anthony Leiserowitz *et al. Politics & Global Warming, December 2018*. Yale University and George Mason University. New Haven, CT: Yale Program on Climate Change Communication, 2019.

S. Lewandowsky, U.K.H. Ecker & J. Cook. 'Beyond Misinformation: Understanding and Coping with the "Post-Truth" Era.' *Journal of Applied Research in Memory and Cognition* 6 (4), 2017.

Gabriel Marcel. *Being and having*. Trans. Katharine Farrer. Westminster, UK: Dacre Press, 1949.

William Nordhaus. Interviewed by A. Smith, Chief Scientific Office of Nobel Media, 2018, https://www.nobelprize.org/prizes/economic-sciences/2018/nordhaus/interview/.

Karl Popper. *The Logic of Scientific Discovery*. London and New York: Routledge Classics, 2002 [1959].

David Roberts. '"Innovation": the latest GOP smokescreen on climate change policy.' *Vox*, 2019, https://www.vox.com/energy-and-environment/2019/1/4/18166400/republicans-climate-change-innovation-policy.

Geoffrey Supran and Naomi Oreskes. 'Assessing ExxonMobil's climate change communications (1977-2014).' *Environmental Research Letters* 12 (8), 2017.

United States Government Accountability Office. *Report to congressional requesters: Climate change: Activities of selected agencies to address potential impact on global migration*, 2019, https://www.gao.gov/assets/700/696513.pdf.

Unknown. 'Smoking and Health Proposal,' Brown & Williamson Records, 1969. Tobacco Industry Influence in Public Policy, Minnesota Documents, https://www.industrydocuments.ucsf.edu/docs/psdw0147.

S.L. van der Linden, A.A. Leiserowitz, G.D. Feinberg & E.W. Maibach. 'The scientific consensus on climate change as a gateway belief: Experimental evidence.' PLoS ONE 10 (2): e0118489, 2015, https://doi.org/10.1371/journal.pone.0118489.

Andrew Watson. 'Only dramatic reductions in energy use will save the world from climate catastrophe: A prophecy?' *Network in Canadian History & Environment*, 2019, http://niche-canada.org/2019/02/27/only-dramatic-reductions-in-energy-use-will-save-the-world-from-climate-catastrophe-a-prophecy/.

Running on Empty by Hafeez Burhan Khan

See the magnificence of Wadi Rum in *Lawrence of Arabia* (1962), *The Martian* (2015) and *Rogue One: A Star Wars Story* (2016). To find out more about Jordan's water shortage see:
https://pangea.stanford.edu/researchgroups/jordan/
https://www.aljazeera.com/indepth/opinion/kushner-deal-century-palestine-jordan-king-abdullah-190522110144692.html
https:/sputniknews.com/middleeast/2019012710
71865008-jordan-israel-water-pipeline/
https://www.usaid.gov/jordan/water-and-wastewater-infrastructure
https://thearabweekly.com/jordan-water-crisis-worsens
-mideast-tensions-slow-action,
https://www.economist.com/middle-east-and-africa/2017/12/02/
jordans-water-crisis-is-made-worse-by-a-feud-with-israel
https://www.independent.co.uk/news/world/middle-east/jordan-water-crisis-israel-relationship-middle-east-peace-security-a8596621.html
https://www.nature.com/news/jordan-seeks-to-become
-an-oasis-of-water-saving-technology-1.22598
Jordan's solar initiatives and green growth plan can be explored by visiting https://www.euronews.com/2019/01/18/
jordan-s-switch-to-renewable-energy-with-solar-power
http://www.jordantimes.com/news/local/green-growth-plan-lead-jordan -more-eco-friendly-sustainable-economy-%E2%80%94-officials

Refugees by Tawseef Khan

For basic data on refugees, see various reports by the UN Refugee Agency (UNHCR) at unhcr.org. On the relationship between Islam and asylum, see Volker Turk, 'Reflections on Asylum and Islam', *Refugee Survey Quarterly* 27 (2) 7-14 2008; and G M Arnaout, 'Asylum in the Arab-Islamic Tradition', Office of the United Nations High Commissioner for Refugees, International Institute of Humanitarian Law, Geneva, 1987. On climate refugees see the reports from Environmental Justice Foundation, *No Place Like Home: where next for climate refugees?* and *The Gathering Storm: Climate*

Change, Security and Conflict (London, May 2019 and March 2014, respectively) – available from ejffoundation.org

See also: Alexander Betts and Paul Collier, *Refuge: Transforming a Broken Refugee System* (Penguin, 2018).

Prosperous Communities by Shanka Mesa Siverio

The reports mentioned in this essay include: Anne Power and Jake Elster, 'Environmental issues and Human Behaviour in Low-Income areas in the UK', Economic and Social Research Council, London, November 2005; Michael Jacobs, 'Green Growth: Economic Theory and Political Discourse', The Centre for Climate Change Economics and Policy (CCCEP) and The Grantham Research Institute on Climate Change and the Environment, LSE, October 2012 (can be downloaded from lse.ac.uk); and Kristen Lyons and David Ssemwogerere, 'Carbon Colonialism: Failure of Green Resources' Carbon Offset Project in Uganda', The Oakland Institute, Oakland, CA, 2017. Oher citations include: Tim Jackson, *Prosperity Without Growth* (Routledge, London, 2011); Kate Raworth, *Doughnut Economics: Seven Ways to Think Like a 21st Century Economist* (Random House, New York, 2018); and Odeh Al Jayyousi, *Islam and Sustainable Development* (Gower, London, 2012).

Babel Bricks by C Scott Jordan

The films mentioned in this article include: nominated for Academy of Motion Picture Arts and Science 2018 Best Foreign Language Film: *Zimna Wojna (Cold War)*, directed by Pawel Pawlikowski, Kino Swiat, Warsaw, 8 June 2018. *Werk Ohne Autor (Never Look Away)*, directed by Florian Henckel von Donnersmarck, Walt Disney Studios Motion Pictures, Burbank, 4 September 2018. *Roma*, directed by Alfonso Cuarón, Espectáculos Fílmicos El Coyúl, Mexico City, 30 August 2018. Manbiki Kazoku *(Shoplifters)*, directed by Hirokazu Kore-eda, GAGA Pictures, Tokyo, 13 May 2018. *Capernaum,* directed by Nadine Labaki, Sony Pictures Classics, New York, 17 May 2018. *Gränis (Border)*, directed by Ali Abbasi, TriArt Film, Stockholm, 10 May 2018. Climate Change Documentaries: *This Changes*

Everything, directed by Avi Lewis, based on the book by Naomi Klein, Abramorama, New York, 13 September 2015. *The Devil We Know,* directed by Stephanie Soechtig, The Film Collaborative, Los Angeles, 21 January 2018. Also discussed: *Interstellar*, directed by Christopher Nolan, Warner Bros. Pictures, Burbank, 26 October, 2016. *Aniara*, directed by Pella Kågerman and Hugo Lilja, Film Constellation, London, 7 September, 2018. *Liúlàng Dìqiú (The Wandering Earth)*, directed by Frant Gwo, China Film Group Corporation, Beijing, 5 February 2019. *Beasts of the Southern Wild*, directed by Benh Zeitlin, Fox Searchlight Pictures, Los Angeles, 20 January 2012. *The House that Jack Built*, directed by Lars von Trier, TrustNordisk, Hvidovre, 14 May 2018. *At Eternity's Gate*, directed by Julian Schnabel, Curzon Artificial Eye, London, 3 September 2018, and *Game of Thrones*, created by David Benioff and D.B. Weiss, Home Box Office, New York, 2012.

Double Exposure by Marjorie Allthorpe-Guyton

The quote from a Biennale panel in 2001 is from *La Biennale di Venezia 2001: An International Round Table June 2001*, Wimbledon School of Art and Nuova Icona, Venice, p.24. The quote for the 2019 UN Global Assessment Report on species extinction is taken from https://www.ipbes.net/news/Media-Release-Global Assessment pp.5,6. All other quotes are from *May You Live In Interesting Times*, Biennale Arte 2019, short guide, frontispiece, p.43, p.41 and p.39, respectively.

See also: Marjorie Allthorpe-Guyton, 'The Aesthetics of Promise' *Critical Muslim 8*: *Men in Islam* (Hurst, London, 2013) p.135-144; and 'Monuments to Hubris' *Critical Muslim 24: Populism* (Hurst, London, 2013) p.153-164

Obituary: Uncle Idris by M Iqbal Asaria

S M Idris's comments on knowledge can be found in Claude Alvares, *Multiversity* (Other India Press, Goa, 2004), p.23, 21. 'The Penang Declaration on Science and Technology' appears as an Appendix in Ziauddin Sardar, *The Revenge of Athena: Science, Exploitation and the Third World*

(Mansell, London, 1988). 'Dissenting Knowledge' pamphlets, edited by Vinay Lal, can be obtained from Amazon.com.

CONTRIBUTORS

Hafsa Abdurrahman, aged 8, lives in Surrey and is a bookworm ● **Mohamed Aidid**, aged 15, is Bristol's Youth Mayor ● **Uzma Ali** is working on her first novel, her debut poetry collection will be published by Waterloo Press in 2020 ● **Marjorie Allthorpe-Guyton** is former President of the International Association of Art Critics (AICA), British Section ● **Iqbal Asaria** is an expert on Islamic economics, finance and banking ● **Farid Bitar** is a well-known Palestinian poet ● **Moiz Bohra** is a doctoral candidate at Imperial College London, where he is researching energy transition pathways for Qatar ● **James Brooks** is a science journalist and writer ● **Canon Giles Goddard** is Vicar of St John's Church in Waterloo, London ● **Christopher B Jones** is a Senior Fellow of the Centre for Postnormal Policy and Futures Studies and faculty member in the Graduate School of Public Policy and Administration, Walden University, Minneapolis ● **C Scott Jordan**, philosopher and futurist, is Executive Assistant Director of Centre for Postnormal Policy and Futures Studies ● **Gazala Khan** grew up in the US and is exploring her Belgian and Bhopali heritage ● **Hafeez Burhan Khan** worked as an archaeologist for English Heritage and Birmingham University before becoming a teacher ● **Tawseef Khan** is a lawyer by profession who is currently writing a book on Muslim identity in the West ● **Ehsan Masood**, science journalist and writer, is Chair of the Muslim Institute ● **Helen Moore**, an award-winning British ecopoet and socially engaged artist, has published two poetry collections ● **Muhammad Akbar Notezai** is Quetta correspondent of daily *Dawn* ● **Yovanka Paquete Perdigao**, a Bissau-Guinean writer, editor and translator, is assistant editor at Dedalus Books ● **Fareha Rahman**, aged 15, is involved in climate activism in Bristol and is a volunteer for Islamic Relief ● **Samia Rahman** is director of the Muslim Institute ● **Shanka Mesa Siverio** runs her own architecture practice in London ● **Gordon Blaine Steffey** is the Barbara Boyle Lemon '57 and William J. Lemon Professor of Religion and Philosophy at Randolph College in Lynchburg, Virginia ● **Medina Tenour Whiteman**, writer, musician, and permaculture enthusiast based in the Alpujarras, is the author of *Huma's Travel Guide to Islamic Spain*.